3rd Annual BMES/FDA Frontiers in Medical Devices Conference 2017

Washington, DC, USA
16 - 17 May 2017

ISBN: 978-1-5108-6458-0

Printed from e-media with permission by:

Curran Associates, Inc.
57 Morehouse Lane
Red Hook, NY 12571

Some format issues inherent in the e-media version may also appear in this print version.

Printed by Curran Associates, Inc. (2018)

Biomedical Engineering Society (BMES)
8201 Corporate Drive
Suite 1125
Landover, MD 20785-2224
USA

Phone: 301-459-1999 / 877-871-2637
Fax: 301-459-2444

membership@bmes.org

Additional copies of this publication are available from:

Curran Associates, Inc.
57 Morehouse Lane
Red Hook, NY 12571 USA
Phone: 845-758-0400
Fax: 845-758-2633
Email: curran@proceedings.com
Web: www.proceedings.com

TABLE OF CONTENTS

PODIUM ABSTRACTS

IDEATION

PRE-CLINICAL

PRODUCT DEVELOPMENT

CLINICAL STUDY

MARKET RELEASE AND POST-MARKET

POSTER ABSTRACTS

IDEATION

PRE-CLINICAL

PRODUCT DEVELOPMENT

CLINICAL STUDY

MARKET RELEASE AND POST-MARKET

Electrodynamic modeling of bacterial biofilm impedance sensing

Ryan C. Huiszoon, Sowmya Subramanian, Thomas E. Winkler, Sam Preza, and Reza Ghodssi
University of Maryland, College Park

Introduction: We present an electrodynamic model of bacterial biofilm to study and evaluate the detection of biofilm growth with impedance-based microsensors. Bacterial biofilms present a ubiquitous problem in human health; they account for 65% of infections and are especially prevalent in hospital-acquired and device-associated infections in the United States [1]. Bacterial biofilms form when planktonic bacteria adhere to a surface, such as a medical implant like a urinary catheter, and encase themselves in an extracellular polysaccharide matrix (ECM). This ECM affords the bacteria protection from traditional antimicrobials, leading to the recurrence and spread of infections [2]. Impedance sensors provide a simple and sensitive method for monitoring biofilm growth on their surface, which allows the early detection needed to prevent infection [3]. The sensor response is affected by a myriad of parameters; the effect of biofilm thickness and electrode spacing on impedance are examined in this work.

Materials and Methods: We utilized a COMSOL 3D electrodynamic simulation to model a biofilm impedance sensor, shown in Figure 1(a). The model consists of a 400 μm x 50 μm x 100 μm (L x W x H) channel with a pair of gold electrodes whose separation is varied from 25 μm to 50 μm to 100 μm. The model then consists of a uniform layer of biofilm – thickness varied from 0 μm to 30 μm – with the remaining volume being the electrolyte lysogeny broth (LB) media. Bacteria are modeled as spherical shells, with an interior of cytosol surrounded by a membrane, embedded in ECM. The overall effective permittivity and conductivity of the biofilm is determined via Maxwell's Mixture Theory assuming a homogenous 10% dispersion of bacteria in the ECM [4,5].

Results and Discussions:

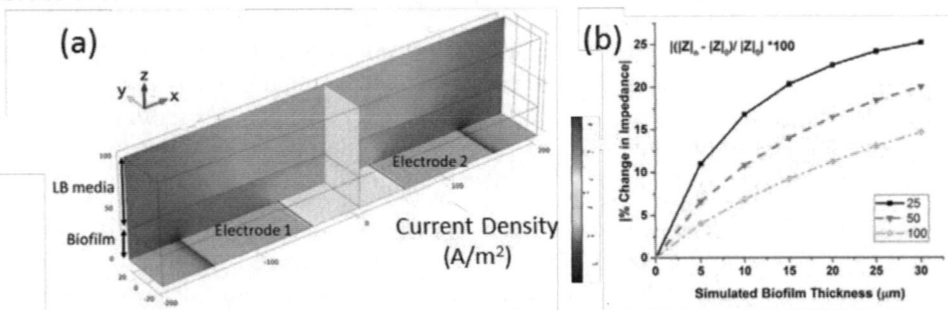

Figure 1: (a) 3D electrodynamic model with 30 μm thick biofilm grown uniformly over the surface with a pair of 100 μm wide electrodes spaced 100 μm apart, showing current density (A/m²) in the x-direction at 100 Hz and 5 mV. (b) Percent magnitude change in impedance at 100 Hz for three different electrode widths and spacings with biofilm.

The model yields the current density at 100 Hz, which was found experimentally to be the most sensitive, shown in Figure 1(a), which determines the system impedance. Simulations were run for three different electrode spacings with biofilm thickness ranging from 0 to 30 μm, and the results are shown in Figure 1(b). Narrower electrode spacing showed higher sensitivity to biofilm, particularly for lower biofilm thickness. However, the smaller spacing also reached signal saturation faster, indicated by the rapidly decreasing slope for 25 μm electrode spacing. The larger spacing displayed a longer linear range.

Translational Impact: Biofilm impedance sensors should be designed in a manner which considers the tradeoff between sensitivity and linear range. The electrodynamic model for the detection of biofilm growth can be extended to evaluate sensor function on more complex implant geometries, such as the cylindrical interior lumen of a catheter. Considering the high complexity of the biofilm composition, the electrodynamic modeling of biofilm should provide a good theoretical background for understanding experimental impedimetric sensing data.

Disclosure Statements: There are no conflicts of interest.

References:

[1] J. W. Costerton et al, *Int J Artif Organs* (2005), **28**(11) p. 1062-1068
[2] J. W. Costerton et al, *Science* (1999), **284**(5418) p. 2043-2046
[3] S. Subramanian et al, *Biomed. Microdevices* (2016), **18**(95)
[4] A. DiBasio et al, *Colloids Surf., B* (2011), **84**(2), p. 433-441
[5] K. Asami et al, *Prog. Polym. Sci.* (2002), **27**(8), p. 1617-1659

A Quantitative Study of the Effect of Wedge Deformity on Thoracic Volume by Virtual Scheuermann's Models

Po-Chih Lee[1], Arthur Erdman[1], Charles Ledonio[2], David Polly[2]
[1]Department of Mechanical Engineering, [2]Department of Orthopaedic Surgery, University of Minnesota

Introduction: Scheuermann's kyphosis (SK) is a sagittal deformity that happens in adolescents. This disease is diagnosed based on the thoracic kyphosis angle (KA) larger than 50 degrees (measured from superior endplate of T1 vertebra to inferior endplate of T12 vertebra) with anterior wedge equal to or larger than 5 degrees in least three consecutive vertebrae. The progressive wedging vertebrae (Fig. 1a) may increase KA and cause shortened thoracic length. Thoracic volume (TV) includes the space inside the rib cage and is bounded by the inner surface of the vertebrae, ribs, costal cartilages and sternum. Severe deformity in SK patients often leads to restrictive lung disease due to smaller TV for lung expansion. Judging from the severity, doctors will determine the treatments: bracing or surgery. Here, we present a process for building five virtual SK patients to show the effect of the wedge deformity (WD) on TV. In future clinical settings, the process can be reproduced to predict the TV changes in SK patients and help doctors decide the timing of intervention.

Materials and Methods: CT-scan data of the rib-cage of a healthy male adult was segmented creating a standard template via Mimics 17.0. This template was used to reconstruct the thorax of five adolescents (three boys and two girls, mean 14 years) with different KA baselines (mean 40.2 degrees) via frontal and lateral X-rays and the reconstruction was implemented in Blender 2.74 with customized add-ons. Also, some assumptions were made based on clinical features of SK disease: 1. At least three consecutive wedging vertebrae possessed wedging angle equal to or larger than 5 degrees. 2. WD above T5 vertebra was not considered. 3. The Apex along the thorax was located at T9 (Fig. 1b). Other assumptions were proposed to simplify the models: 1. Global and local rib cage deformities only occur in sagittal plane. 2. Sternum and ribs do not locally deform when WD progressively bends the thoracic spine. 3. All consecutive endplates remained parallel to each other while WD happens. 4. KA for virtual SK models started with 60 degrees with three WDs and every 10 degrees more of progression would add one more wedging vertebra and so on. Finally, each rib-cage was morphed to simulate SK models with different KAs in five virtual patients and our developed program was capable of creating and measuring the TV in Blender.

Results and Discussion: In Fig. 1c, an average of reduction from 60-degree KA to 110-degree KA was 18.5% in five virtual SK patients. With the increased wedging angle and increased number of wedged vertebrae, the thoracic KA progressively increased, and the TV decreased in an approximate linear fashion by a mean of 3.5% for every 10 degrees more of progression in the five virtual SK patients. The virtual SK patients showed that increased wedging number and wedging angle not only affected the local and global spine deformity but also reduced the TV for lung expansion in adolescent SK patients if no intervention was made.

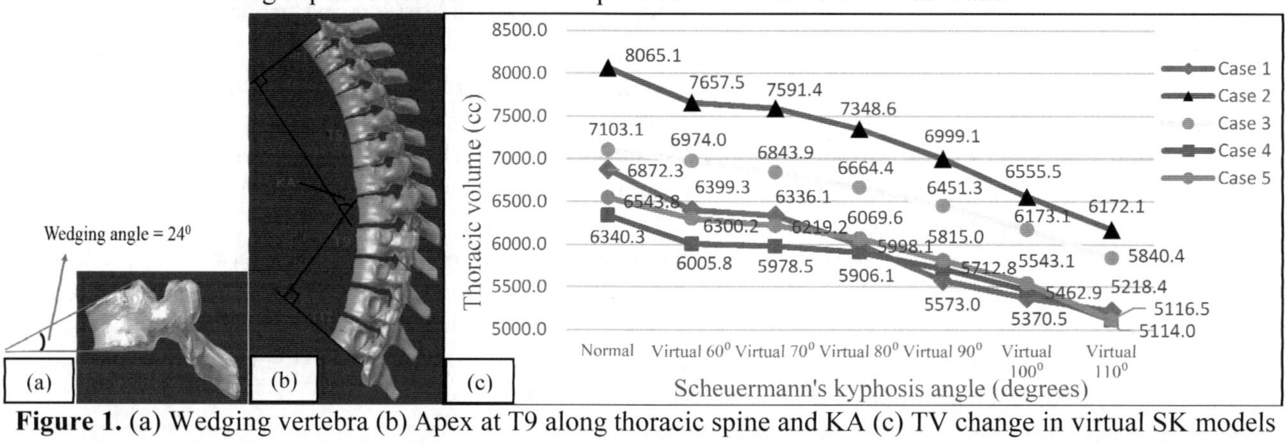

Figure 1. (a) Wedging vertebra (b) Apex at T9 along thoracic spine and KA (c) TV change in virtual SK models

Translational Impact: TV can be a vital parameter to judge the pulmonary function in spine patients. Five virtual SK patients with different KAs were used to investigate the effect of the WD on the spine deformity and the TV variations. The process of constructing the virtual SK patients can be utilized in future clinical settings to predict and track the TV decrease based on the WD along the thoracic spine. Doctors can decide about timing for intervention for SK patients based on KA with TV variations as a guide.

Disclosure Statement: All authors have no conflicts of interest to declare.

Optimizing an Endoscopic Bioimpedance-Based Surgical Margin Assessment Device

Ryan Halter, Aditya Mahara, Ethan Murphy
Thayer School of Engineering, Dartmouth College, Hanover, NH, USA

Introduction: Positive surgical margins (PSMs) following prostate cancer surgery increase the chance of cancer recurrence. Unfortunately, there are no approaches to intraoperatively assess margin status, and instead surgeons need to rely on post-operative histological assessment of the resected prostate. We have developed endoscopic electrical impedance sensing probes for intraoperative surgical margin assessment[1]. The electrical impedance of prostate cancer has been shown to be significantly different than benign prostate and peri-prostatic tissues. Here we explore through simulations the design of an optimized multi-electrode probe for use in intraoperative surgical margin assessment.

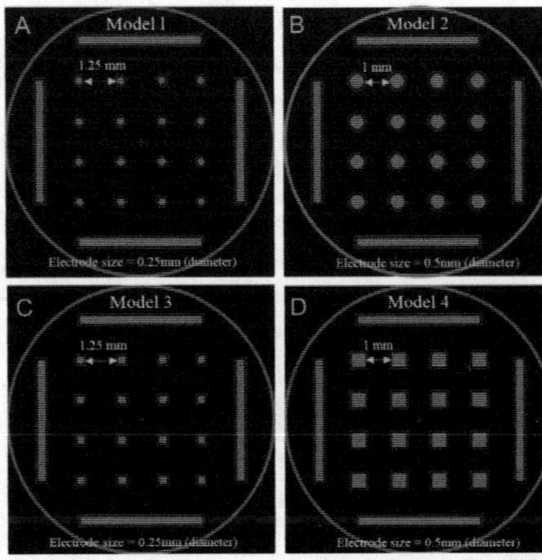

Fig. 1 Electrode array models explored

Materials and Methods: A simulation study was conducted using four different electrode models (Fig. 1). Model 1 and 2 include a grid of 4x4 circular voltage sensing (VS) electrodes with width 0.25 and 0.5 mm, respectively, fitting within a 10 mm diameter probe face. Model 3 and 4 include a grid of 4x4 square VS electrodes with width 0.25 and 0.5 mm, respectively. The spacing between the electrodes in model 1 and model 3 is 1.25 mm from one edge of a VS electrode to the other, and 1 mm between electrodes for model 2 and model 4. Therefore, model 1 and 3 will provide resolution up to 1.25 mm and model 2 and 4 up to 1 mm. The spaces between the electrodes diminish with increasing electrode size. A denser electrode array or use of electrodes with larger diameter was not possible due to trace width limitations on PCB fabrication. Four tetrapolar patterns are investigated for each array (i.e. current driven between two electrodes and voltage sensed between two other electrodes). A custom Matlab-based electromagnetic model is used to compute resistance, for the four pattern sets assuming a 0.1 S/m conductivity background. The mesh used in the simulation study has 30747 nodes and 20 electrodes (Fig. 2).

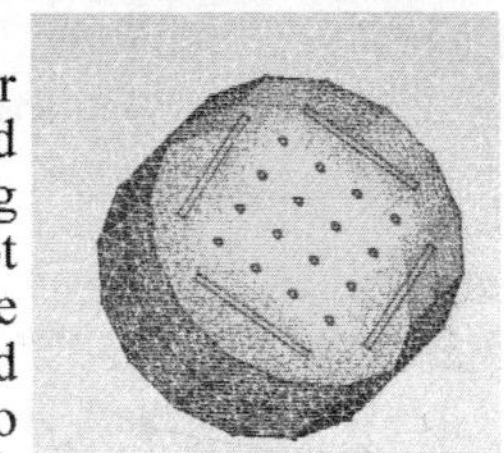

Fig. 2 Example FEM Mesh

Results and Discussion: Figure 3 shows simulated resistances for the four patterns explored. Pattern set 1 includes VS electrodes that are at equipotential; as expected the impedance values are low (~0). Pattern set 2 includes VS electrodes that are not equipotential, but are orthogonal to the current drive electrodes, resulting in relatively low impedances. Pattern sets 3 and 4 have VS electrodes in between current electrodes and in line with the direction of current flow; these patterns are expected to result in relatively higher impedances compared to pattern set 2. Pattern set 4 results in highest impedance values because both VS electrodes are closest to the current electrodes and furthest apart from each other. *Importantly, the variation between models is not significant.* Therefore, in order to maximize electrode size (to minimize contact impedance) and avoid potential charge build up at the corners of square electrodes, model 2 is selected as optimal for fabrication. A prototype device with this electrode configuration was fabricated (Fig 4).

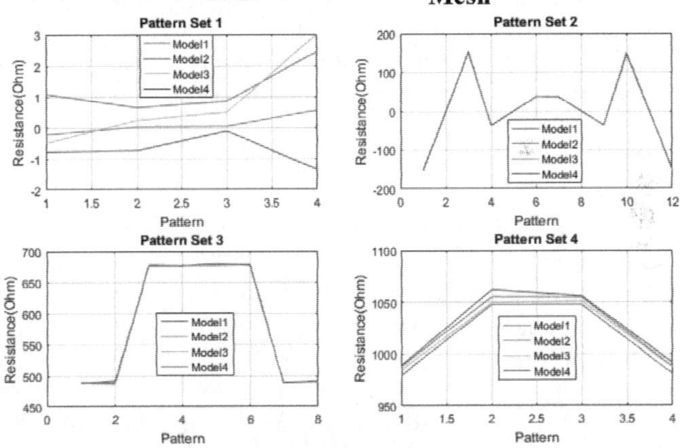

Fig. 3 Example resistances of multiple patterns and models explored

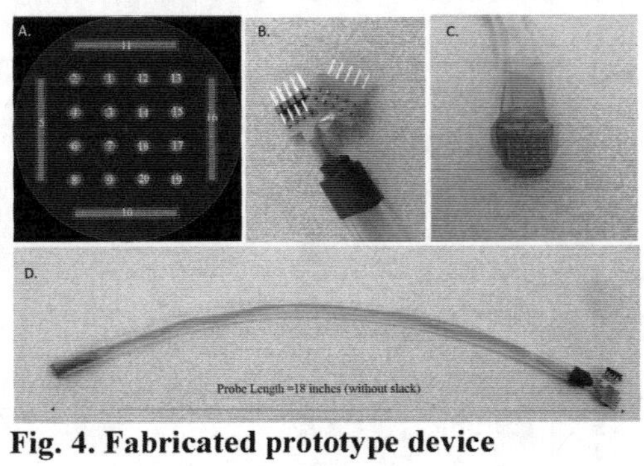

Fig. 4. Fabricated prototype device

Translational Impact: An optimized electrical impedance sensing probe for intraoperative surgical margin assessment will help to reduce the incidence of PSMs and ultimately improve long-term outcomes for men following radical prostatectomy.

Disclosure Statement: No conflicts of interest to report

Acknowledgements: This research is supported by NIH/NCI grant number 1RO1CA14302001A1.

References: 1. Mahara A et al, IEEE TMI, 34(7), 1590-1601, 2015

Biomechanical Effects of Femoral Component Axial Rotation in Total Knee Arthroplasty

Mohammad Kia, PhD, Timothy M Wright, PhD, Michael B Cross, MD, David J Mayman, MD, Andrew D Pearle, MD, Peter K Sculco, MD, Geoffrey H Westrich, MD, Carl W Imhauser, PhD
Hospital for Special Surgery, New York, NY

Introduction: The correct amount of external rotation of the femoral component during total knee arthroplasty (TKA) is controversial because the resulting changes in biomechanical knee function are not well understood. Therefore, we used a computational knee model to address the following research questions: (1) Does external rotation consistently unload the medial collateral ligament (MCL) as is clinically expected and what is the effect on lateral collateral ligament (LCL) tension? (2) What are the consequences of external rotation on the contact loads across the tibiofemoral joint and on tibiofemoral kinematics?

Materials and Methods: We incorporated a posterior stabilized TKA (Logic, Exactech, FL) into a previously-developed computational knee model [1] in three unique sets of tibiofemoral geometries (Fig. 1). Implant geometries were virtually implanted using hip to ankle computed tomography scans of three cadaveric knees using measured resection and anterior referencing surgical techniques. Collateral ligament properties (stiffness and slack length) were standardized across each knee model [1-3]. The femoral component was externally rotated from 0 to 9° relative to the posterior condylar axis in 3° increments. At each increment, the knee was flexed under 500 N compression from 0 to 90°. The computational model predicted collateral ligament forces, compartmental contact forces, and tibiofemoral internal/external and varus-valgus rotation through the flexion range.

Results and Discussion: The computational model predicted that femoral component external rotation unloads the MCL between 30° and 90° of flexion (Fig. 2A-C). When the femoral component was placed in 9° of external rotation rather than 0°, the maximum force carried by the MCL decreased a respective 55, 88, and 297 N in knees one, two, and three. LCL tension increased only with extreme external rotation (6 to 9°) depending on the knee. The amount of offloading with external rotation was inconsistent from knee-to-knee. Increasing femoral component external rotation from 0 to 9° led to a maximum decrease in medial contact forces of 90, 190, and 570 N in knees one, two, and three, respectively. External rotation of the femoral component led to increased external tibial rotation in early flexion and increased varus alignment in late flexion corroborating previous findings [4].

Translational Impact: While the model intuitively confirms that external rotation of the femoral component will decrease MCL and medial compartment loads, the effect was variable across each knee. Other factors such as the location of the transepicdonylar axis, femoral geometry, selected implant size and placement may effect tibio-femoral loads and contribute to the variability observed in this study. In addition, the conventionally used 3° of external rotation relative to the posterior condylar axis combined with anterior referencing does not produce consistent changes in knee biomechanics suggesting alternate referencing systems should be considered in total knee arthroplasty.

Disclosure Statement: Timothy Wright and Geoffrey Westrich receive royalties from Exactech.

References: [1] Kia M et al. *J Biomech Eng* 2016. [2] Robinson JR et al. *J Biomech* 2005. [3] Wilson WT et al. *J Orthop Sports Phys Ther* 2012. [4]Miller MC et al. *Clin Orthop Relat Res* 2001.

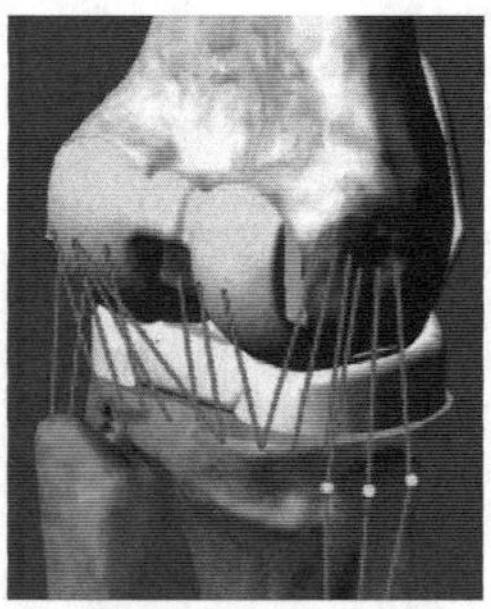

Fig1: TKA knee model

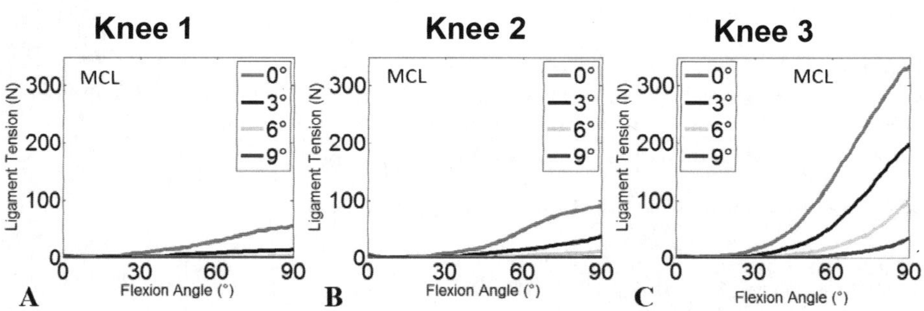

Fig 2: Medial collateral ligament tensions predicted by the three knee models

Leaflet Geometry and Anisotropy of Bioprosthetic Heart Valves: Do They Matter?

Rana Zakerzadeh[1], Will Zhang[1], Michael C. H. Wu[2], Fei Xu[2], Ming-Chen Hsu[2], Keefe Manning[3], Michael Sacks[1]

[1] Institute for Computational Engineering and Science, Department of Biomedical Engineering, The University of Texas at Austin
[2] Department of Mechanical Engineering, Iowa State University
[3] Department of Mechanical Engineering, Pennsylvania State University

Introduction: The most popular replacement heart valves continue to be bioprosthetic heart valves (BHV), which are fabricated from biologically derived biomaterials. While these devices benefit from low thrombogenicity and excellent hemodynamics similar to the native valve, device failure continues to be the result of leaflet structural deterioration mediated by fatigue and tissue mineralization. Moreover, valve performance is highly dependent on leaflet geometry. There is thus a profound need to enhance the physical realism of the BHV dynamics simulation in the design context. In this study, we developed a framework for parametric study of BHV to investigate the influence of the geometry of the leaflets on heart valve performance and to simulate the effect of leaflet material model of the leaflets on valve dynamics.

Materials and Methods: We modeled the leaflets as a thin shell structure using different constitutive models. In particular, both nonlinear isotropic and anisotropic material models have been considered and compared. A Rhino-Grasshopper based interactive geometry design platform is used to construct the valve according to the aortic root geometry. We simulate the valve deformations with various perturbations of the tissue properties including the collagenous fiber orientation, as well as the fiber and matrix stiffness to investigate the effect of alteration of the tissue mechanical properties. We also examine the effect of changing leaflet's shape on valve dynamics by adjusting some geometrical parameters such as leaflet's thickness and surface curvature. To approximate this problem, we use a thin shell isogeometric structure dynamic finite element solver for parametric studies of the material properties by applying a physiological transvalvular pressure load to the leaflets to enforce the pressure difference between left ventricle and aorta. Parametric studies of the leaflet's geometry have been considered in the framework of an immersogeometric fluid-structure interaction (FSI) methodology. Validation of the FSI model is performed using pulsatile duplicator heart valve system to compare the FSI simulation results to the experimental data. The results of this validation make a convincing case for the fidelity of the numerical method for our analysis.

Results and Discussion: The functionality of BHV was evaluated by its effective orifice area and leaflet coaptation without regurgitation. The influence of the geometry on valve dynamics has been performed by comparing the valve effective orifice area and coaptation area for different designs of the leaflets. We observed a noticeable influence of the surface curvature on the effective orifice area of the valve and the overall tissue deformation. Sensitivity results for different values of fiber stiffness show that perturbing fiber stiffness coefficient in our anisotropic material model has a significant influence on the valve deformation only at the closing stage. The reason is that in dynamic simulation of heart valve in a full cardiac cycle, the opening phase is in the small strain regime, when fibers has not been recruited and the tissue matrix dominates the valve deformation. Therefore, the role of the anisotropic term is only pronounced during the closing stage, that representing the fiber action.

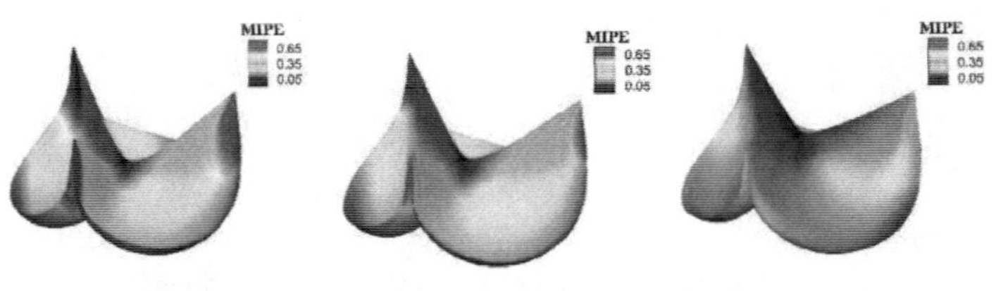

Figure 1. Deformation of the valve colored by Maximum in-plane principal Green-Lagrange strain (MIPE) for different values of fiber stiffness parameter. The chosen time point is at the end of cardiac cycle when the valve is fully closed.

Translational Impact: The goal is to present a framework which enables us to use more physically realistic soft tissue material model in heart valve simulations and to identify the role of leaflet's shape on valve functionality, indicated by the effective orifice area during the opening and the coaptation area during the closing. These results can provide guidelines for designing leaflet tissues to improve valve durability. We are currently formulating a phenomenological material model that covers the interaction between the tissue's collagen fibers and its ground matrix as well as the interaction between different fibers and permanent set. This computationally tractable material model facilitates the integration of an anisotropic model into our FSI solver.

Computer-assisted Design of Microfluidics

Junchao Wang and William H. Grover
Department of Bioengineering, Bourns College of Engineering, University of California, Riverside, 92521, USA

Introduction: Since the emergence of the first lab-on-a-chip devices in the late 1970s, microfluidic chips have found a variety of applications in biomedical engineering and clinics. But while the range of possible applications for microfluidics has blossomed, the process of designing microfluidic chips has remained relatively unchanged since the 1970s. Researchers still draw new microfluidic chips designs by hand, fabricate and test them, and repeat this process until a functional device design is found. This trial-and-error process is expensive and inefficient, and it slows the spread of microfluidic technologies to new fields. In this presentation, we will share our recent work developing computer algorithms and software tools that enable users to design microfluidic chips much more efficiently.

Materials and Methods: In our work, we were motivated by the question, *is it possible to create functional microfluidic devices without actually designing them?* We answered this question in two different ways: **Random design of microfluidics**: We wrote a custom MATLAB program that generated thousands of random microfluidic chip designs, simulated the behavior of each design using COMSOL Multiphysics, and stored the results in a publicly-accessible database (http://random.groverlab.org). Users can query this database to find a chip design suitable for a given application. As a proof of concept, we created a library of 10,513 fluid mixer chips that can generate any three arbitrary concentrations at the same time. Sixteen of these designs were etched in glass, anodically bonded to silicon, and tested using food coloring and cell growth media to confirm that the designs function as predicted [1]. **Microfluidics-optimized particle simulation algorithm (MOPSA)**: Many microfluidic chips that we would like to design algorithmically will contain particles like cells and microbeads, but the particle simulation tools in existing commercial software are largely not suitable for predicting the behavior of particles microfluidic chips. To address this need, we developed a microfluidics-optimized particle simulation algorithm (MOPSA) that simulates particle trajectories with greater accuracy than existing commercial software. MOPSA treats particles as two- or three-dimensional objects (not single points) and uses an empirical correction to make particle-wall interactions more lifelike. MOPSA is implemented in MATLAB and uses a fluid velocity profile solved by COMSOL Multiphysics.

Results and Discussion: Figure 1 shows one of the 10,513 different random microfluidic mixers that we generated and simulated. The chip was fabricated, tested, and found to generate the expected fluid concentration. Figure 2 shows the predicted trajectories of three different-sized cells flowing through a typical triangle-post deterministic lateral displacement (DLD) sorter chip. MOPSA correctly predicted the paths of the different cells while the commercial software COMSOL Multiphysics failed.

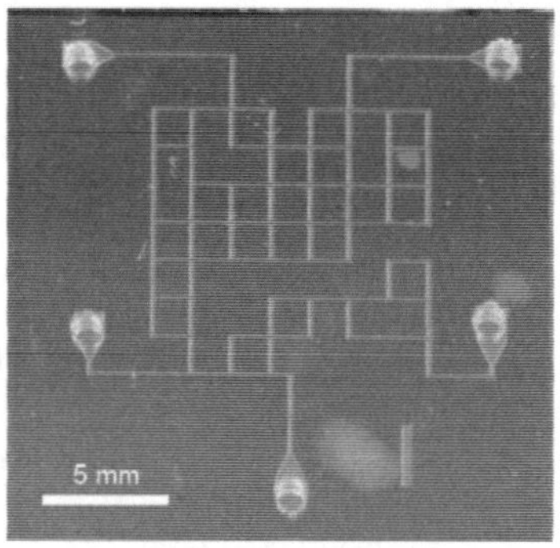

Figure 1: A randomly-designed microfluidic mixer chip capable of delivering three desired concentrations of a solute.

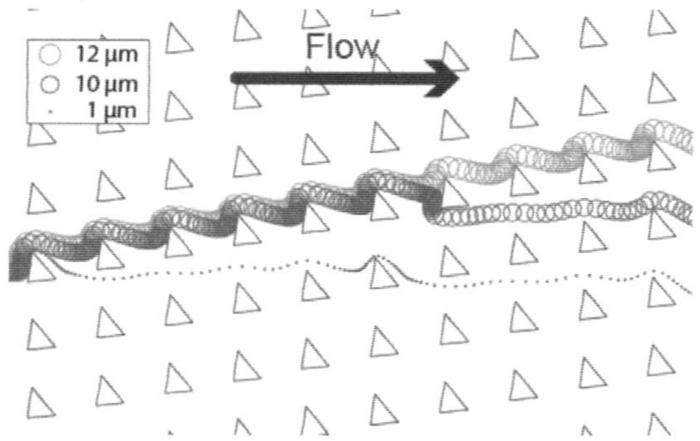

Figure 2: MOPSA-predicted paths of three different-sized cells through a DLD sorter chip.

Translational Impact: As libraries of random chip designs proliferate in the future, even complex lab on a chip devices like cell sorters could be created in seconds by users with no experience in microfluidics.

Disclosure Statement: No conflicts of interest to disclose.

References: 1. J. Wang, P. Brisk, and William H. Grover. *Lab on a Chip* 16 (21), 4212-4219.

Optimizing the Cardiac Defibrillation Stimulus

Anton Shpak, Socrates Dokos
Graduate School of Biomedical Engineering, University of New South Wales,
Sydney, 2052, Australia

Introduction: The success of external cardiac defibrillation depends largely on the waveshape and amplitude of the applied electric stimulus. The aim of our study was to develop a simple computational model to perform a comprehensive comparison of different waveforms used in modern defibrillators, thereby gaining insights into the optimal defibrillating impulse waveshape.

Materials and Methods: A simple 3D model of defibrillation was implemented using COMSOL Multiphysics finite-element software to simulate re-entrant arrhythmia in the heart, with defibrillation shocks applied to electrodes on the chest. The waveforms selected for simulated defibrillation were Rectangular (Mono- and Biphasic), Exponential (Mono-, Bi- and Triphasic), Rectilinear Biphasic and Quasi-sinusoidal current pulses. The efficacy criterion for assessing defibrillation was the defibrillation threshold energy, defined as the minimal energy for a given current stimulus waveshape to successfully defibrillate the heart.

Results and Discussion: Results are summarized in table 1. In general, Biphasic waveforms were more efficient than their Monophasic counterparts. The Rectilinear Biphasic waveform was superior to all other waveforms for which the phase 2 amplitude was 0.7 of phase 1, and phase 1 duration was 0.35 of the total waveform duration. This was followed by the Exponential Biphasic, Triphasic, and Rectangular Biphasic (equal phase amplitude) waveforms. The least effective were the Quasi-sinusoidal, the Exponential Monophasic and Rectangular Monophasic waveforms. Waveforms with short first phase durations tended to defibrillate more efficiently, along with exponential impulses with lower tilts. The second phase of the Biphasic waveforms exhibited a threshold, below which the waveform ceased to defibrillate, suggesting that this second phase is critical in ensuring that the heart's electrical activity is effectively reset. Our results also indicated that the most effective waveforms tended to follow charge-balanced biphasic configurations.

Waveform	DTE (mJ)	Optimal Waveform Parameters
Rectangular Monophasic	2.977	A = 25 mA, D = 75 ms
Rectangular Biphasic (equal phase amplitude)	0.698	A = 20 mA, D = 27.5 ms, D1F = 0.75
Rectangular Biphasic (phase 2 amplitude 0.7 of phase 1)	0.637	A = 20 mA, D = 37.5 ms, D1F = 0.35
Exponential Monophasic	0.850	A = 20 mA, D = 90 ms, Tilt = 0.7
Exponential Biphasic	0.667	A = 25 mA, D = 37.5 ms, D1F = 0.35, Tilt = 0.6
Exponential Triphasic	0.704	A = 25 mA, D = 40 ms, D1F = 0.25, D2F = 0.5, Tilt = 0.6
Quasi-sinusoidal (phase 2 amplitude 0.5 of phase 1)	0.827	A = 35 mA, D = 35 ms, D1F = 0.5

Table 1. Simulation results of Defibrillation Threshold Energy (DTE) for various waveforms. Waveform parameters represent Amplitude (A), Total Waveform Duration (D), Phase 1 duration as a fraction of total waveform duration (D1F), Phase 2 duration as a fraction of total waveform duration (D2F), Tilt = $(I_{start}-I_{end})/I_{start}$, where I_{start} and I_{end} are the absolute current amplitudes at the start and end of the waveform respectively.

Translational Impact: The simulation of various waveforms and their modalities employing computational modelling made it possible to estimate the defibrillation energy requirements and compare efficacy of these waveforms. The results of this study could be used to determine optimal parameter combinations and tailor waveforms in order to achieve maximal efficacy in defibrillation.

Disclosure Statement: The authors declare that no conflicts of interest were associated with this study.

Local Specific Absorption Rate (SAR) in computational models of blood vessel compared to ASTM phantom

Kyoko Fujimoto[1], Leonardo M. Angelone[1], Elena Lucano[2], Peter Serano[1], Sunder S. Rajan[1], Maria I. Iacono[1]
[1]U.S. Food and Drug Administration, Center for Devices and Radiological Health, Silver Spring, MD
[2]Sapienza University of Rome, Department of Information Engineering, Electronics, Telecommunication, Italy

Introduction: An implanted stent can cause heating of surrounding tissues during a Magnetic Resonance Imaging (MRI) scan. As a part of premarket submissions, stents are tested based on standards such as ASTM F2182 − 11a [1]. The heating effect measured is proportional to the local specific absorption rate (SAR). However, the local SAR *in-vivo* can vary significantly with the positioning in the MRI bore, location in the body and coil characteristics. In addition, the local SAR in arteries or veins relative to surrounding regions has not been evaluated due to the coarse spatial resolution (>2mm) currently used to discretize computational human models which is not sufficient to model the geometry of the vessels. In this study, we compared the SAR inside the ASTM phantom and in a high-resolution human model with four different imaging landmarks, and showed that *in-vivo* exposure in specific anatomical locations is significantly higher than that in the ASTM phantom.

Material and Methods: Sim4Life v3.0.2 (Zurich Med Tech, Zurich, Switzerland) based on the finite-difference time-domain (FDTD) method was utilized. The birdcage coil was modeled based on the MITS1.5 radiofrequency coil (Zurich Med Tech, Zurich, Switzerland) [2]. The high-resolution AustinMan model [3] with $1mm^3$ isotropic spatial resolution and 64 anatomical structures was converted to be compatible to the simulation software by matching each anatomical structure with one of the available 55 anatomical structures in Sim4Life. The blood vessels (Fig. 1 in red) were discretized with 1mm grid and the rest of the model with 2mm grid. Five simulations were performed at 64MHz: one simulation with the ASTM phantom and four simulations with the AustinMan at four imaging landmarks (brain, heart, hip, and knee) as in Fig. 1. SAR maps were then obtained by averaging over 0.1g, 1g and 10g of the surrounding tissues. All the simulations were normalized to obtain an averaged whole-body SAR of 2 W/kg.

Results and Discussion: The maximum 1g SAR value of the ASTM phantom was 8.6 W/kg. Blood vessels (white arrows in Fig.2) and fat had different SAR values thus were easily discernable, whereas the SAR values of blood vessels (purple arrows in Fig.2) and muscle were similar. Fig.3(a) shows that the femoral vessels with the hip bone landmark showed 3-fold higher local SAR (up to 23 W/kg) compared to the ASTM. Similarly, the popliteal vessels with the knee landmark resulted in 5-fold higher local SAR (up to 40 W/kg) compared to the ASTM phantom (Fig.3(b)).

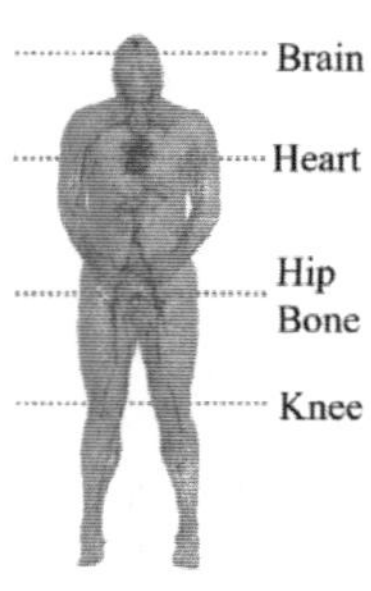

Fig.1 High-resolution AustinMan with four imaging landmarks.

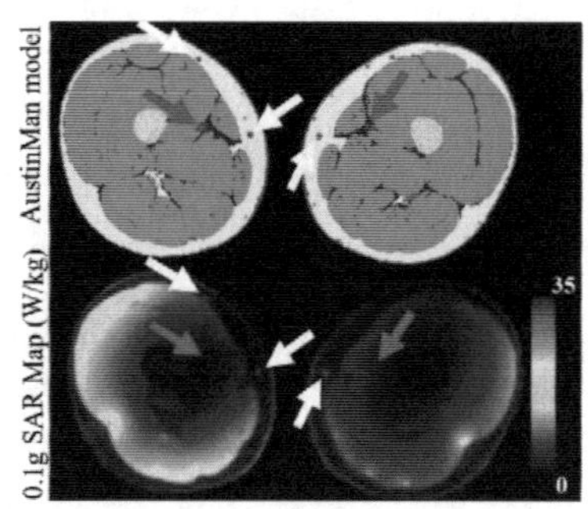

Fig.2 Segmentation of AustinMan and the 0.1g SAR map of the knee landmark in the thigh region.

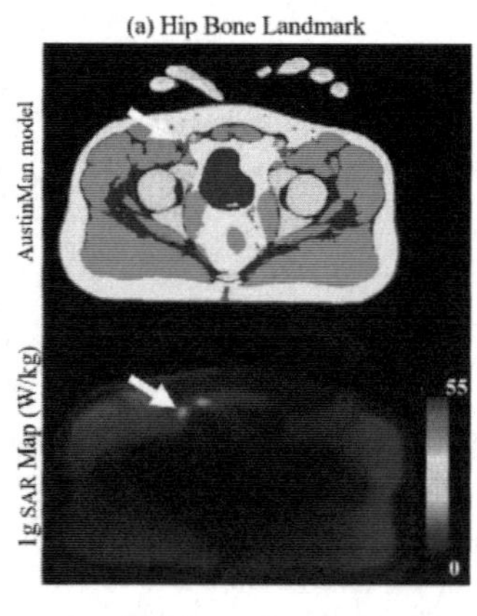

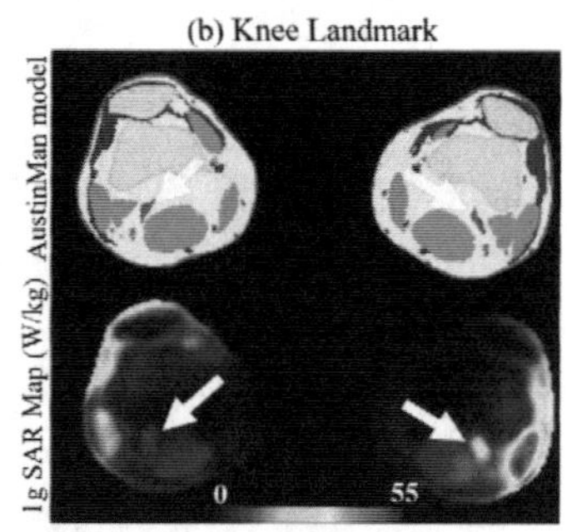

Fig.3 Arteries with high SAR values at the hip bone landmark (a) and the knee landmark (b).

Translational Impact: At specific imaging landmarks, the value of SAR can be significantly higher than that of the ASTM phantom. Thus, testing based on the ASTM phantom may need to be rescaled to better reflect the *in-vivo* local SAR.

Acknowledgement and Disclaimer: The mention of commercial products, their sources, or their use in connection with material reported herein is not to be construed as either an actual or implied endorsement of such products by the Department of Health and Human Services.

References: [1] ASTM (2011): F2182-11a. [2] Lucano, E. IEEE Trans. Biomed. Eng. (2016):63(8):1591-1601 [3] Massey, J.W. Proc. Bioelectromagnetics Soc. 34th Annu. Meeting (2016): 3346-3349.

Modeling The Role of Surface Interactions in Initiation of Thrombosis

Liudi Zhang[1], PhD, Brendan Casey[3] , PhD, Dennis Galanakis[2], MD, Clement Marmorat[1], Shelby Skoog[3],PhD
Katherine Vorvolakos[3],PhD, Marcia Simon[4],PhD, Miriam H. Rafailovich[1], PhD
[1]Department of Materials Science and Engineering, Stony Brook University, NY, 11794
[2]Department of Pathology, Stony Brook University, NY, 11794
[3]Food and Drug Administration, Center for Devices and Radiological Health, Silver Spring, MD, 20903
[4]Department of Oral Biology and Pathology, School of Dental Medicine, Stony Brook University, 11794

Introduction:

Thrombosis is a well-known side effect of the introduction of foreign materials into the bloodstream, as might exist in medical devices including but not limited to stents, valves, and intravascular catheters. Despite many reported studies, the body's response to foreign materials in contact with the blood remains poorly understood. Current preventive methods consist of drug eluting coatings on the devices or the systemic administration of standard anticoagulants. Here we present a potential mechanism by which surface chemistry can affects fibrinogen conformation and thus affects platelet adhesion and consequently thrombus formation. Our findings suggest a possible coating which enables endothelial cell adhesion while preventing platelet adhesion.

Materials and Methods

Polymers commonly used in medical devices were spun cast onto silicon wafer and exposed to protease-free human fibrinogen under both static and flow condition. The polymers were also spun cast onto Kapton sheets and exposed to circulating platelet rich plasma.

Results and Discussion

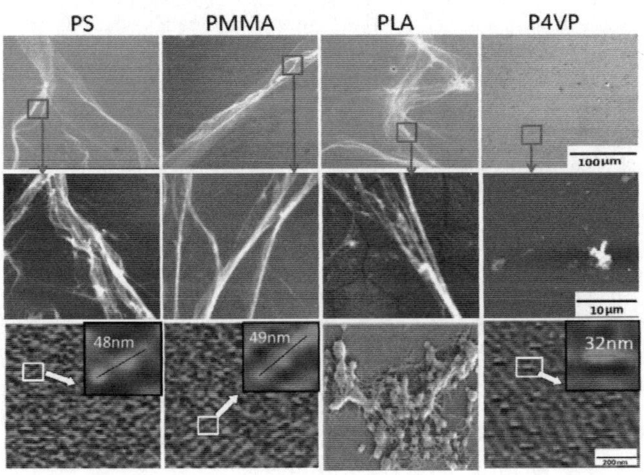

Figure 1. Upper panel: optical microscope images of the PS, PMMA, PLA, and P4VP polymer coated substrates incubated with 4mg/mL fibrinogen, showing fiber networks yielded only on hydrophobic surfaces. **Central panel:** SPM topographical images of fibrinogen fibers formed on a PS substrate. Note that the fiber is composed of smaller fibers, attached to the surface, which are intertwined and form very large surface anchored rope like structures. **Lower panel:** High resolution (lateral force mode SPM) images of the fibrinogen coating obtain after incubation with 0.1mg/mL fibrinogen solution . The left two inserts show a tri-nodular fibrinogen molecule (indicated by the arrows) on the indicated PS and PMMA (hydrophobic) surfaces. By contrast, the P4VP (hydrophilic) surface displays variably globular and oblong images one of which is shown similarly magnified. Platelets, which adhere to the fibers, are shown on PLA in the center image.

We observed that on hydrophilic surfaces, fibrinogen is adsorbed via α-C regions, while the γ400-411 platelet-binding dodecapeptide on the D region becomes exposed, and fibrinogen fibers do not form. In contrast, fibrinogen is adsorbed on hydrophobic surfaces via the relatively hydrophobic D and E regions, exposing the α- C regions while rendering the γ400-411 dodecapeptide inaccessible. Fibrinogen adsorbed on hydrophobic surfaces is thus able to recruit other fibrinogen molecules, especially in the form of soluble fibrin, through α -C regions and polymerize into large fibrinogen fibers, similar to those formed *in vivo* in the presence of thrombin (figure 1). Moreover, the γ400-411 dodecapeptide is available only on the large fibers, while elsewhere throughout on the hydrophobic surface, the α -C regions are exposed. When these surfaces were exposed to gel-sieved platelets or platelet rich plasma, a uniform monolayer of platelets, which appeared to be activated, was observed on the hydrophilic surfaces. In contrast, large agglomerates of platelets were clustered on fibers on the hydrophobic surfaces, resembling small nucleating thrombi. On hydrophobic surfaces, with only a monomeric coat of fibrinogen, only endothelial cells were adsorbed from a culture of mixed platelets and endothelial cells.

Translational Impact: The ability to model the conformation of fibrinogen on different surfaces enables one to predict whether platelet or endothelial cell binding occurs, minimizing non-enzymatic thrombogenesis.

Acknowledgement: This work was supported by the NSF-CBET FDA/SIR # 1239560 and by the Food and Drug Administration's Center for Devices and Radiological Health. *Authors have no conflict of interest.*

Towards Validation of a Physiological Model Intended for Design & Evaluation of Automated Fluid Resuscitation Systems

Bahram Parvinian[1], Ramin Bighamian[2], Pras Pathmanathan[1], Tina Morrison[1], George Kramer [3], Christopher G. Scully[1], Jin-Oh Hahn [2]

1. Office of Science and Engineering Laboratories, Center for Devices and Radiological Health Food and Drug Administration Silver Spring MD 20993
2. Department of Mechanical Engineering, University of Maryland College Park MD 20742, USA
3. Department of Anesthesiology, University of Texas Medical Branch, Galveston, TX 77555, USA

Introduction: Critical care therapies such as fluid resuscitation of hemorrhagic patients have the potential to be enhanced by system automation. This is mainly due to the labor intensive nature of these therapies as well as understaffing in critical care units, particularly in areas with a public health emergency (e.g., earthquake). There is a significant need to identify performance metrics, stability, robustness, and overall potential failure modes of automated fluid resuscitation devices to assess safety and effectiveness. Model-based strategies for design and evaluation of fluid resuscitation controllers provide a viable approach. However, in order for this method to provide meaningful results validity of fluid resuscitation model should be established. In this presentation we will present validation steps and results for a fluid resuscitation model amenable to controller design and evaluation.

Materials and Methods: A model of hemodynamic response to fluid loss and infusion was constructed by integrating three sub-models of blood volume response, cardiac output, and blood pressure response. The main modeling philosophy was to keep the model as simple as possible (i.e. fewest number of parameters) without having appreciable loss in accuracy for closed-loop controller design and non-clinical (*in silico*) evaluation context of use. The parameters of the model were tuned using an optimization scheme that calculated the optimal parameters of the three sub-models that would minimize the error between model prediction and experimental data. The parameters were obtained from an animal study with conscious and splenectomzied sheep[1]. The identified parameters were examined for physiological relevance and were subjected to a sensitivity analysis, which allowed for relatively insensitive parameters to be identified. Population averaging and covariate analysis were also conducted to assess the influence of subject specific weight and initial blood volume on average model parameters. The following validation activities were then applied to the candidate averaged model: 1) Assessment of model structure and qualitative evaluation of model's physiological relevance 2) Internal validation[2] using a subset of experimental data from the same animal study that did not take part in parameter identification. Model performance was quantified by calculating the Root Mean Squared Error (RMSE) between the model output and experimental results.

Results and Conclusions: The model demonstrated high degree of physiological relevance and transparency. Calibration results indicated that the model form and structure follows the physiology of fluid transfer between plasma and interstitial fluid compartments. The agreement of the model output for mean arterial pressure (the endpoint for closed-loop controller) with experimental results for individualized parameters was 9.8%±2.6(%RMSE). Covariate analysis did not appreciably improve the accuracy of model output. Model agreement with experimental data deteriorated when the average model was compared to the internal validation data set (RMSE= 16.8%±5.1%). This deterioration was contributed to small data set (n=4) and significant inter-subject variability.

Translational Impact: In this study we present a model of fluid resuscitation amenable to design and evaluation of closed-loop fluid resuscitation systems. Validation of such models has not been adequately undertaken in current scientific literature. We presented initial steps for validating this model for design and evaluation of fluid resuscitation controllers. This context of use is highly relevant to regulatory applications. Future research directions include enhancing the model, increasing the number of animal subjects with homogenous physiological response, and applying an external validation[2] data independent of the training data set.

Disclosure Statement: Ramin Bighamian and Jin-Oh Hahn have a pending patent related to the blood volume model presented in this study.

References: [1] Kramer, G. C., et al. (2017). "Inaccuracy of Urine Output Measurements due to Urinary Retention in Catheterized Patients in the Burn ICU." J Burn Care Res **38**(1): e409-e417. [2] Guidance for Industry"Population Pharmacokinetics" Food and Drug Administration Center For Drug Evaluation and Research 1999.

A Novel Method in Predicting Aneurysmal Hemodynamics after Endovascular Coiling

Hooman Yadollahi-Farsani[1], Marcus Herrmann[1], Brian Chong[2, 3], David Frakes[2]

[1]School for Engineering of Matter, Transport and Energy Arizona State University Tempe, AZ, USA

[2]School of Biological and Health Systems Engineering Arizona State University Tempe, AZ, USA

[3]Mayo Clinic Hospital Phoenix, AZ, USA

Introduction: A brain aneurysm is a bulging of a weakened blood vessel wall which if ruptured, leads to high mortality rates [1]. In order to fix an unruptured aneurysm, the gold standard approach suggests the use of coils [1]. In this approach, the aneurysmal sac is filled with coils to reduce the blood activity inside the sac and hence, easing the clotting process. A numerical replication of this procedure can be significantly important in treatment planning. The common approach uses a software package to firstly, deploy the coils, secondly, mesh the fluid domain excluding the coils and, lastly solve the appropriate fluid flow equations. However, since coils are made of very tiny struts, meshing their boundary layers requires a large number of mesh elements. This makes the fluid flow simulations extremely resource intensive and time consuming. In an attempt to bypass the aforementioned issue, the coiled region is assumed to be a homogeneous porous region [2]. However, there is a significant drawback associated with this assumption. The complexity of the deployed coil geometry in the aneurysm invalidates the homogeneity assumption. Herein, we introduce a novel method which considers the coiled region as a heterogeneous porous domain. Accordingly, while keeping the decreased computational costs associated with the porous domain assumption, the accuracy is not sacrificed by the homogeneity assumption. This study is expected to play an important role in facilitating pre/post treatment hemodynamic studies as well as the design and manufacturing of endovascular coils.

Materials and Methods: The geometry of the aneurysm was constructed using 3D rotational angiography images and coils were deployed within the aneurysmal sac. The deployed coils' geometries were then used to generate a map of porosity. Specifically, the coiled region was broken into a lattice of hexahedra; the porosity was found subsequently within each hexahedral by calculating the volume of the coil portion. The generated map served as an input to the volume-averaged Navier-Stokes equations [3] to solve for the fluid flow domain. The fluid domain was discretized into tetrahedra and then the independency of the solution from the mesh resolution was confirmed. The flow solver CLIFF was used as the incompressible Navier-Stokes solver.

Results and Discussion: We used root mean squared velocity (VRMS) as a quantity to investigate the blood activity within the aneurysmal sac. We observed that the deployment of coils within the aneurysm can reduce the VRMS by up to 130%. This is in accordance with the initial intention of using coils to reduce the aneurysmal blood activity. Employing the homogeneity assumption, however, predicted only 30% reduction in aneurysmal VRMS. This proved the deficiencies which are associated with the latter assumption. Finally, it was observed that a relatively coarse porosity map, in this case 16 hexahedra in each direction, can recover the coiled aneurysm VRMS reduction of 130% within a negligible error band.

Translational Impact: In this study, we developed a novel method which significantly simplifies the blood flow simulations in the presence of endovascular devices. This method can act as an efficient and reliable tool in treatment planning. Physicians will be able to predict the outcomes associated with each treatment option within a relatively short amount of time. Moreover, biomedical companies which are involved in design and manufacturing of endovascular devices will be able to assess their designs with a significant reduction in costs associated with computational resources.

References:

[1] P. Nair *et al.*, *J. Biomech. Eng.*, Jan. 2016, vol. 138, no. 2, pp. 021011–021011.
[2] N. M. P. Kakalis, A. P. Mitsos, J. V. Byrne, and Y. Ventikos, *IEEE Trans. Med. Imaging*, Jun. 2008, vol. 27, no. 6, pp. 814–824.
[3] S. Whitaker, *Transp. Porous Media*, Mar. 1986, vol. 1, no. 1, pp. 3–25.

Evaluation of DBS Implant Interaction with Brain Tissue Through Finite Element Simulation

Logan E. Miller[1], Jillian E. Urban[1], Vincent Whelan[2], Walt W. Baxter[2], Stephen B. Tatter[1], Joel D. Stitzel[1]
[1]Wake Forest School of Medicine, Winston-Salem, NC 27101
[2]Medtronic, Minneapolis, MN 55432

Introduction: Deep brain stimulation (DBS) is used to treat movement disorders such as Parkinson's disease by implanting a lead with electrodes that sends rapid electrical impulses to specific targets in the brain [1]. This technique is an improvement over previous therapies because it temporarily disables overactive target cells instead of destroying them. The DBS technique requires the lead be chronically implanted so stimulation can be constantly applied to the target cells. An important step in evaluating new device designs is examining how the implant interacts with surrounding brain tissue and how its material properties affect this behavior. The current study aims to answer this question through a collaborative effort between industry, academia, and physicians that was established as the result of knowledge sharing at a prior BMES/FDA Frontiers in Medical Device Conference. Finite element (FE) analysis was used to investigate relative motion of the DBS lead in a suite of cases that represent activities of daily living and range in severity. Head kinematics were applied to the skull of a head FE model as boundary conditions and internal interactions and deformation of the lead with respect to the brain tissue and skull/anchor were explored.

Materials and Methods: A high-resolution, anatomically accurate brain FE model, the Atlas-Based Brain Model (ABM), was used to investigate how a typical DBS implant interacts with the brain. The ABM (Figure 1) was derived from the International Consortium for Brain Mapping (ICBM) brain atlas [2]. A series of beam elements representing the DBS lead were placed in the ABM. The location of the lead was determined by creating a common reference frame between the FE model and Stereotaxy Simulator and using typical target coordinates for the subthalamic nucleus (STN). The target location was verified using a probabilistically derived atlas map of the STN. The lead was modeled with beam elements to allow the lead elements to move with the motion of the brain. The lead was attached to surrounding brain tissue using a contact algorithm that ties each node to the nearest element face. Linear and angular velocity boundary conditions from an experimentally-derived, moderately severe, impact case (C383-T1, Hardy et al. (2001)) were applied to the ABM to drive the simulation [3]. Lead resultant linear motion relative to initial lead location was determined with respect to a reference line collinear with the lead at time=0 and aligned with the motion of the skull throughout the simulation.

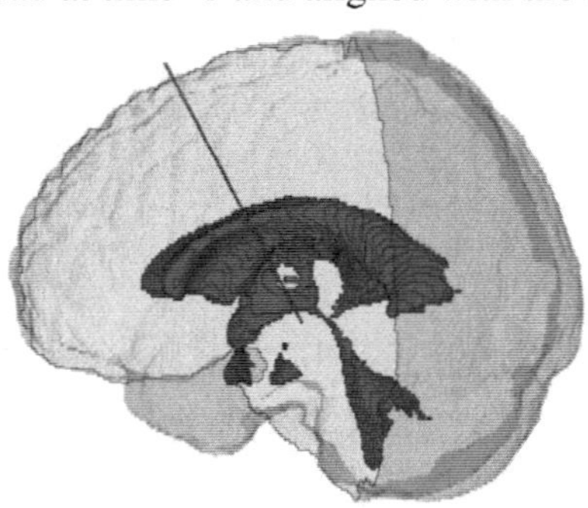

Figure 1. The DBS lead inserted into the ABM.

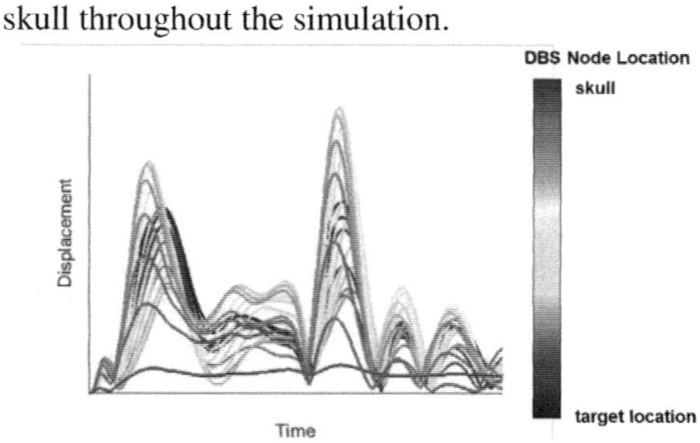

Figure 2. Resultant relative displacement of DBS nodes.

Results: Relative displacements of the DBS nodes are displayed in Figure 2 for a frontal head impact of moderate severity. The maximum displacement occurred 56 mm (70% of intracranial lead length) from the target location.

Conclusions: The results of this study highlight the importance of collaboration between academia, industry, and physicians to improve product development. The response of an implanted DBS lead was examined during a frontal impact of moderate severity. The methodology presented in this study can be used to investigate the lead's behavior in a wide range of scenarios including daily activities such as running or walking down steps.

Disclosure Statement: The authors have no conflicts of interest to disclose.

References: [1] Obeso, J.A., et. al., 2001 [2] Miller, L.E., et. al., 2016 [3] Hardy, W.N., et. al., 2001

A Numerical Study on the Feasibility of the Deep Brain Stimulation (DBS) Surgical Navigation based on Scalp Electric Potential Recordings

Seyed Reza Atefi[1], Maria Ida Iacono[2], Leonardo Angelone[2] and Giorgio Bonmassar[1]
[1]Department of Radiology, Massachusetts General Hospital, Harvard Medical School, USA
[2]U.S. Food and Drug Administration

Introduction: Patients suffering depression, chronic pain or neurological conditions such as Parkinson's disease can benefit from Deep Brain Stimulation (DBS) of certain regions of the brain. Precise placement of DBS electrodes is key for the optimal clinical outcome of the DBS surgical procedure. However, localizing the exact DBS target position using conventional imaging techniques such as MRI and CT during surgery is still a great challenge. Thus, we propose using electroencephalography (EEG) to localize the position of the electrodes based on the patient specific MRI and a 60μs pulse generated on a DBS electrode pair. This study aims to preliminarily validate such navigation technique by performing anatomically and geometrically realistic numerical simulations and to estimate the signal intensities needed to achieve a robust DBS localization and navigation during surgery.

Materials and Methods: A high resolution (1 mm^3) head model and a DBS model were used to simulate the Electric Field generated by DBS in subthalamic nucleus (STN) of the brain. Simulation was done in Finite Differences Time Domain (FDTD) and the Electrical Field generated by the DBS was transferred into Multiphysics (COMSOL, Burlington MA) to calculate the potential distribution on the scalp. Potential distribution on the scalp was then sampled at 256 locations and according to 10-20 EEG electrode placement system. Electric potential samples were then fed into the dynamic Statistical Parametric Maps (dSPM) algorithm (Brainstorm,) to solve the inverse problem, and localize the DBS. Once DBS location is estimated it will be compared to its actual location in the brain.

Results and Discussion: The electric potential on the scalp followed a dipole pattern oriented according to the DBS electrodes axis in the head (**Figure 1a-b**). The 256 sampled scalp potentials (**Figure 1c**), allowed localization of the DBS via dSPM (**Figure 1d**) Although localization results are proximal to the STN, further simulations are needed to evaluate the accuracy of the DBS localization algorithm for different electrode pairs and targets. Furthermore, simulations are also needed to compare the performance of the proposed method in the presence of noise, using reduced electrode numbers (16,32, 64 and 128 electrodes) and different localization algorithms such a aLORETA

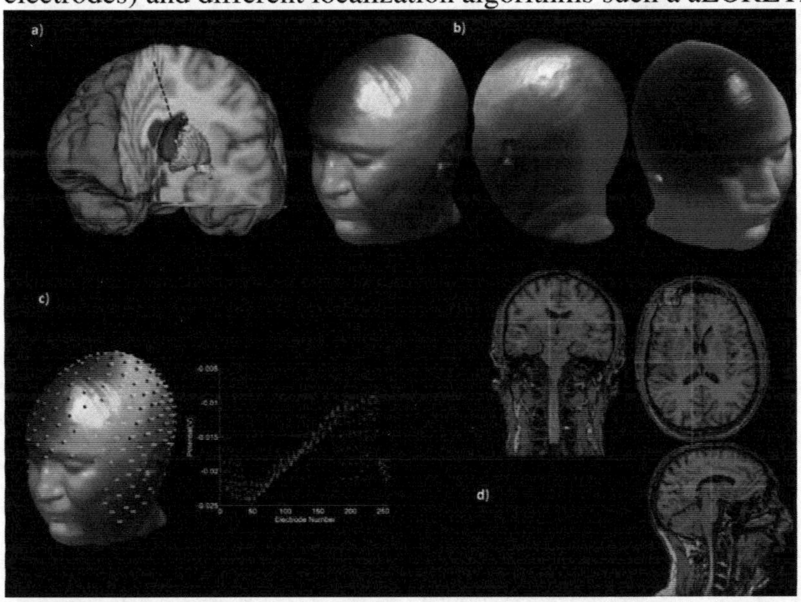

Figure 1. a) Illustration of the DBS placed in the STN **b)** DBS potential distribution on the scalp **c)** 256 scalp potential samples ranging between -0.025-0.0005 V **d)** localization of the DBS based on the 256 scalp potential samples

Conclusions: These results shows proof of principle that it is possible to localize the subcortical DBS source based on its potential distribution on the scalp generated by a DBS pulse, which normally constitutes an artefact for EEG.

Conflicts of Interest: The Authors declare that there is no conflict of interest. The mention of commercial products, their sources, or their use in connection with material reported herein is not to be construed as either an actual or implied endorsement of such products by the Department of Health and Human Services

Valve Interstitial Cell Phenotypic State After Surgical Repair: An Integrated Experimental-Computational Approach

Salma Ayoub[1], Chung-Hao Lee[2], Samuel T. Potter[1], Jordan L. Graves[1], and Michael S. Sacks[1]
[1] Center for Cardiovascular Simulation, Institute for Computational Engineering and Sciences and Department of Biomedical Engineering, The University of Texas at Austin. [2] School of Aerospace and Mechanical Engineering, The University of Oklahoma

Introduction: Despite the phenomenal level of reliability of heart valves, more than five million people are diagnosed with valvular disease in the United States each year, with approximately 95,000 annular valve replacement surgeries and 20,000 mortalities per year [1]. Though ring annuloplasty for mitral regurgitation is beneficial in the short-term, it has been shown to be less promising in the long term, with repair failure as high as 60% [2-3]. Mechanical stress is a strong etiological factor: alterations in mechanical loading caused by surgical repair lead to stress-induced changes in mitral valve interstitial cell (MVIC) function that affect both tissue structure and composition, ultimately leading to repair failure. In this work, we develop a robust experimental-computational approach that allows us to answer specific questions on mitral valve (MV) mechanobiology and provides insight into the micromechanics of surgically repaired valves. This work is thus three-fold: the first part focuses on the design and development of a planar biaxial bioreactor system. The second part involves the use of an established finite element model of the mitral valve [4] to estimate *in vitro* pre-stresses and functional stresses before and after different surgical repair scenarios. We then use the biaxial bioreactor system to apply the estimated stresses to stimulate MV samples and follow-up with downstream biological assessments of the valves.

Materials and Methods: In brief, a three-stack cyclic stretch biaxial bioreactor was developed to replicate the complex physiological biaxial strains of the MV (Fig. 1A) The design consists of four linear actuators and two load cells (one on each axis). A four-pillared stacking system in the middle of the systems holds the three chambers made of wear-resistant and temperature-stable polysulfone (Fig. 1B). A pin attachment system is used to mount the tissue inside the specimen chambers (Fig. 1C). The system can be used as either a strain- or stress-controlled device and allows the operator to modulate strain and membrane tension along each axis.

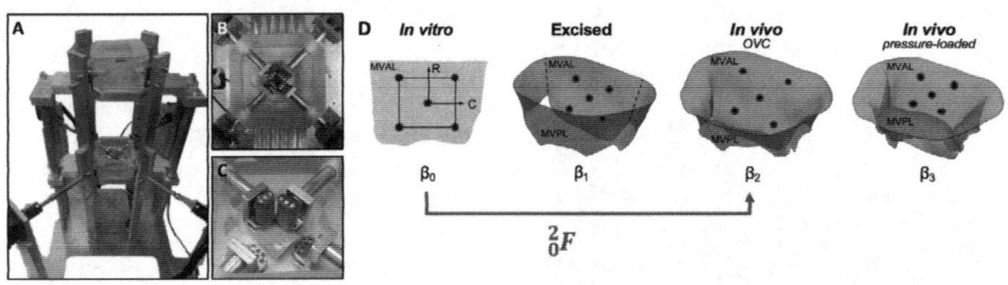

Figure 1. (A) Completed biaxial bioreactor system. (B) Top view of specimen chamber, (C) Tissue attachment system (tissue phantom in blue). (D) 4 key kinematic states that bridge the gap between *in vitro* and *in vivo* reference configurations.

Results and Discussion: This integrated experimental-computational approach allows us to emulate *in vivo* loading conditions and thus *in vivo* remodeling events in our *in vitro* system. We will bridge the gap between the *in vivo* functional and *in vitro* experimental configurations by applying pre-stresses estimated from the inverse modeling framework (30.4 kPa and 20.5 kPa in the circumferential and radial directions, respectively) to all tissue samples prior to the cyclic stretch treatment.

Translational Impact: To develop robust tools that can help improve surgical repair, we must first understand the underlying mechanisms of MV compensatory response after ring annuloplasty by linking tissue-level stresses to MVIC mechanobiological response. A wide array of experimental studies have captured phenomena that drive VIC homeostasis—the current challenge lies in transforming this wealth of information into a holistic understanding of how cell response dictates long-term disease and surgical repair failure. The integrated experimental-computational approach that we present in this work will serve as a platform for identifying biomechanical treatments that optimize surgical repair and elucidate mechanisms of disease progression.

Disclosure Statement: The authors have no conflicts of interest to disclose.

Acknowledgements: National Institute of Health grant R01HL119297 to MSS and American Heart Association Pre-Doctoral Fellowship 17PRE33420135 to SA and Scientist Development Grant 16SDG27760143 to CHL. The authors would like to acknowledge Annie Mara, Brenda Rodriguez-Nino, and Lindsay Lehman.

References: [1] Go et al. *Circulation*, 129:228-92, 2014. [2] Flameng et al., *Circulation*, 107: 1609-13, 2003. [3] Gillinov et al., *J Thorac Cardiovas Surg*, 135:885-93, 2008. [4] Lee et al. *Biomech Model Mechanobiol,* in press.

Towards Accurate Simulations of the Patient-Specific Mitral Valve Annuloplasty Repair

Andrew Drach [1], Amir H. Khalighi [1,2], Robert C. Gorman [3], Joseph H. Gorman III [3], and Michael S. Sacks [1,2,4]
[1] Center for Cardiovascular Simulation, Institute for Computational Engineering and Sciences, The University of Texas at Austin
[2] Department of Mechanical Engineering, The University of Texas at Austin
[3] Gorman Cardiovascular Research Group, University of Pennsylvania
[4] Department of Biomedical Engineering, The University of Texas at Austin

Introduction: Mitral Valve Regurgitation (MVR) is the most prevalent valvular heart disease, characterized by incompetent function of the Mitral Valve (MV). MVR affects at least 300,000 Americans and is present in over 50% of patients with reduced left ventricular function induced by the myocardial infarction (MI). While the repair (annuloplasty and leaflet augmentation) is considered as the preferred treatment strategy, almost a third of the patients experience the recurrence of MVR within 12 months of the treatment. Strong predictive tools are needed to improve the repair outcomes. We propose a robust pipeline to build high-fidelity, attribute-rich finite-element models of the complete mitral heart valve from imaging data. We rely on in-vitro patient-specific data for model development and validation of its predictive capabilities.

Materials and Methods: Five ovine mitral valve samples were imaged using micro-computed tomography in a left-heart in-vitro simulator. We simulated three scenarios: "healthy", "diseased" (myocardial infarction), and "repaired" (flat-ring repair). The specimens were imaged in fully open (unloaded) and fully closed (pressurized to 100 mmHg) states. We used the data from the "healthy" state to build and calibrate the finite element model. We validated the model using the "diseased" and "repaired" datasets. This model was used to simulate the biomechanical response of five MVs to several repair scenarios, including the annuloplasty (AP) with Physio (Edwards Lifesciences, Irvine, CA, USA) ring.

Results and Discussion: The attribute-rich finite element model, and the predictions for the four different scenarios are presented in Figure 1. The surface plots represent the measure of the total collagen fiber recruitment (TFR) in the leaflets in the normalized range from 0 (slack collagen bundles) to 1 (fully straightened/recruited collagen bundles).

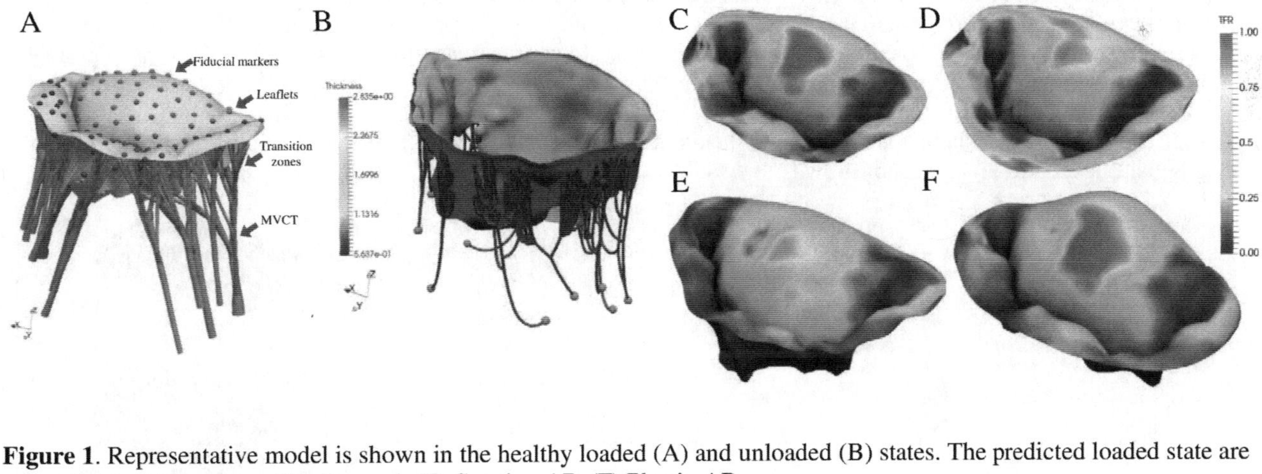

Figure 1. Representative model is shown in the healthy loaded (A) and unloaded (B) states. The predicted loaded state are presented for (C) healthy, (D) diseased, (E) flat-ring AP, (F) Physio AP.

Translational impact: We demonstrated the predictive capability of the model via thorough validation of the predictions for five ovine MVs. The model utility and flexibility was illustrated by performing patient-specific simulations of the AP repair with a several ring shapes. The proposed model can be directly used for medical device design and surgical planning. We believe that the proposed approach will lead to the better outcomes of the AP procedures, and eventually, to the optimal patient-specific AP ring designs. In the future, we plan to extend this work to allow using in-vivo imaging data instead of the in-vitro datasets.

Disclosure statement: The authors declare no conflicts of interest.

Interlaboratory Simulations of Compression-Bending Testing of Spinal Rods

Srinidhi Nagaraja[1], Andrew Baumann[1] Galyna Loughran[2], Gunjan Verma[3], Jason Inzana[2], Maureen Dreher[1], Anup Gandhi[2], and Marc Horner[3]
[1]U.S. Food and Drug Administration
[2]Zimmer Biomet Spine
[3]ANSYS, Inc.

Introduction: ASTM standardized test methods are used extensively to evaluate the mechanical performance of spinal devices, particularly to support their regulatory clearance. By contrast, finite element analysis (FEA) is primarily performed during spinal device development and is infrequently used as part of the regulatory submission. In this study, experimental results for ASTM F1717 (Standard Test Methods for Spinal Implant Constructs in a Vertebrectomy Model) were compared to interlaboratory computational simulations of the test method. In order to develop best practices for computational modeling of this test method, we investigated the influence of several modeling input parameters on the predicted compression-bending response.

Materials and Methods: ASTM F1717 testing was performed on a generic pedicle screw construct (n=4) consisting of monoaxial pedicle screws (Ti-6Al-4V), set screws (Ti-6Al-4V), spinal rods (Stainless Steel 303), and test blocks (UHMWPE) in compression-bending. An interlaboratory FEA study replicating the experimental testing was performed by FDA, ANSYS, and Zimmer Biomet (ZB). Each lab was blinded to experimental results. Material properties were obtained from the literature for Ti-6Al-4V and UHMWPE components. Stainless steel spinal rods were modeled as an isotropic material with multilinear isotropic hardening using material properties determined from tensile testing of dogbone specimens. All interfaces in the spinal construct were modeled with perfectly bonded contact in the initial simulations. After initial simulations were compared to experimental results, model parameters (e.g., element type, discretization, contact conditions) were altered to understand the influence of these parameters on the mechanical behavior of the construct.

Results and Discussion: Interlaboratory simulations of dogbone tensile testing using the determined material properties resulted in excellent agreement with experiment (Fig. 1(a)). ASTM F1717 simulations exhibited less agreement with experiments and greater interlaboratory variability (Fig. 1(b)). Simulation and experimental force-displacement curves diverged after 1mm of displacement. All simulations exhibited stiffer elastic behavior until yield (2% offset method). Simulation yield points did not correlate to experiments. Plastic behavior (displacement greater than 7mm) was also dissimilar between laboratories, with two laboratories displaying softer behavior than experimental constructs. The maximum difference between simulations and experiment was 28%, occurring at 4.4mm displacement. After initial simulations, ZB found that including frictional contact at the rod-screw interface improved agreement with experiment to 7mm of displacement (Fig. 1(b)). Altering element shape (tetrahedral/hexahedral), shape function (linear/quadratic), hardening (isotropic/kinematic), and compatibility had negligible effect on results (data not shown).

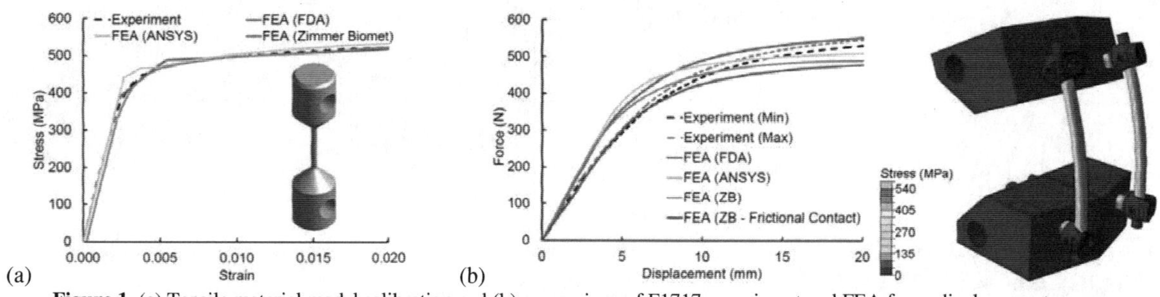

Figure 1. (a) Tensile material model calibration and (b) comparison of F1717 experiment and FEA force-displacement response.

Translational Impact: To increase the use of modeling in orthopedic regulatory submissions, it is critical to establish a standardized framework for computational stress analysis. This study is working towards that goal by outlining the influence of various model parameters on the mechanical response of pedicle screw-rod constructs. Ongoing uncertainty analyses will further elucidate best practices for conducting FEA for this product area.

Disclosure Statement: nothing to disclose

Numerical Optimization of Scaffold Properties for Tissue Engineered Venous Grafts

Jason M. Szafron[1], Ramak Khosravi[1], Christopher K. Breuer[2], Alison L. Marsden[3], Jay D. Humphrey[1]
1 - Department of Biomedical Engineering, Yale University, New Haven, USA. 2 - Tissue Engineering Program and Surgical Research, Nationwide Children's Hospital, Columbus, USA. 3 - Department of Pediatrics (Cardiology), Bioengineering and ICME, Stanford University, Stanford, USA.

Introduction: Tissue engineered vascular grafts have shown great promise as off-the-shelf conduits for use in cardiovascular surgery, particularly in pediatrics and palliative procedures such as Fontan. Nevertheless, these implants exhibit long-term complications due to excessive matrix production and consequent narrowing of the vessel [1]. Recent advances in material science and fabrication techniques allow greater control of scaffold geometries and properties for a greater range of materials. Because of this versatility, computational biomechanical models have been developed to predict growth and remodeling (G&R) outcomes for particular combinations of scaffold properties [2]. While parametric studies are illustrative and increase intuition as to improved designs, truly optimal scaffold designs require identification of quantitative metrics for comparison of outcomes and the coupling of growth and remodeling simulations with an optimization framework. The Surrogate Management Framework (SMF) has been shown to be an effective parameter estimation and optimization method in previous cardiovascular studies [3]. The goal of the coupled simulations presented herein was to identify potentially optimal parameter values from a predetermined set, including scaffold porosity, siffness, and mass degradation rate, which play key roles in the body's response to implanted constructs [4].

Methods: A detailed review of the G&R framework for a cylindrical tube has been described previously [2]. In brief, the graft constituents, including polymer, collagen and smooth muscle cells, were modeled using constrained mixture theory. Rates of inflammation-mediated tissue production and degradation were prescribed based on phenomenological relationships with the scaffold microstructure. The polymer was assumed to have a sigmoidal degradation profile described by a rate-like parameter and an offset. Additionally, mechano-mediated tissue production was assumed to depend on differences in circumferential stress and wall shear stress from homeostatic values for the native vein. The SMF interfaced with iterative G&R simulations to determine optimal parameter values with an objective function based on the difference in linearized stiffness between the graft and native vein. The objective function acts as a quantitative marker for graft performance, and linearized stiffness of vessels has been shown as a key marker of vascular health [5].

Results and Discussion: The coupled simulations revealed the importance of initial parameter choice on graft outcome, while also indicating preferable parameter sets for future experimental testing. Due to the low pressure within the venous circulation, polymers with stiffness values on the order of 1 MPa were predicted to prevent graft dilatation upon loading while limiting mechano-mediated matrix production due to circumferential stresses on the order of those experienced in the native vein. Scaffold porosities of 50% prevented excessive production of inflammatory related matrix components, which have higher stiffnesses than normal matrix materials. Rapid degradation of the polymer, on the order of 2 weeks, was determined to reduce the duration of the inflammatory response and prevent excessive graft thickening. Together, these results highlight the importance of controlling inflammation to prevent stiffening and the associated narrowing of venous vascular grafts.

Translational Impact: This coupled biomechanical – optimization modeling technique reduces the need for costly and time consuming animal procedures to eliminate poor graft designs, especially when scaffold parameters are initially far from optimal. These principles can be applied in similar tissue engineering design problems to reduce the experimental burden and accelerate new device development.

Disclosure Statement: Gunze Limited provided funding for experiments related to this work.

References: [1] N. Hibino et al. "Late-term results of TEVGs in humans..." *JTCVS* 139.2 (2010): 431-436.
[2] K.S. Miller et al. "A hypothesis-driven parametric study of effects..." *Acta Biomaterialia* 11 (2015): 283-294.
[3] A. Ramachandra et al. "Computational simulation of the adaptive..." *J Biomech Eng* 137.3 (2015): 031009.
[4] K. Garg et al. "Macrophage functional polarization (M1/M2) in..." *Biomaterials* 34.18 (2013): 4439-4451.
[5] J. Ferruzzi et al. "Biomechanical phenotyping of central arteries..." *Annl Biomed Eng* 41.7 (2013): 1311-1330.

Development and Full Body Validation of a 5th Percentile Female Finite Element Model

Matthew L. Davis[1,2], Bharath Koya[1,2], Jeremy M. Schap[1,2], F. Scott Gayzik[1,2]
1. Wake Forest School of Medicine
2. Virginia Tech-Wake Forest Center for Injury Biomechanics

Introduction: Computational human body modeling has become an important tool for the development of safety devices in the automotive and defense industries. Such models are often developed to represent a 50th percentile male (M50). However, in order to address the effects of size and sex-related geometrical changes, there is interest in developing models of other cohorts. This study focuses on the female in the 5th percentile of height and weight (F05). As part of the Global Human Body Models Consortium (GHBMC) project, comprehensive medical image and anthropometrical data of the F05 were acquired for the specific purpose of finite element model (FEM) development [1]. The objective of this study is to apply these data to generate and validate a detailed FEM of the F05 in a seated posture representing a vehicle occupant.

Materials and Methods: The F05 occupant (F05-O) model geometry was developed using an image dataset consisting of CT, MRI and upright MRI images of one representative small female subject in the seated, standing, and supine postures. The CAD geometries assembled from the medical images were utilized for mesh development of the F05-O FEM. Structured hexahedral mesh was used in the majority of the body with the exception of the abdomen. In general, material definitions were applied based on GHBMC average male model. The model was compared to experimental data in 10 validation cases ranging from localized rigid hub impacts to full body sled cases. In order to make direct comparisons to experimental data, which represent the mass of an average male, the model was compared to experimental corridors using two methods: 1) post-hoc scaling the outputs from the baseline F05-O model and 2) geometrically morphing the model to the body habitus of the average male to allow direct comparisons. This second step required running the morphed full body model in all cases for a total of 20 simulations. Model outputs were objectively evaluated using ISO/TS 18571.

Results: The F05-O model consists of 981 parts, 2.6 million elements, and 1.4 million nodes (Figure 1) [1]. Thresholds were placed on several element quality criteria: Jacobian (>0.3 for solid elements and >0.4 for shells), tet-collapse (>0.2 for all elements), zero intersections, and a minimum time step of 0.1 μs. Contacts were minimized via node-to-node connections and element property assignment based on the CAD data. Overall, geometrically morphing the model was found to more closely match the target data with an average ISO score for the rigid impacts of 0.76 compared to 0.67 for the scaled responses. Based on these data, the morphed model was then used for model validation in the vehicle sled cases and attained an average weighted score of 0.69.

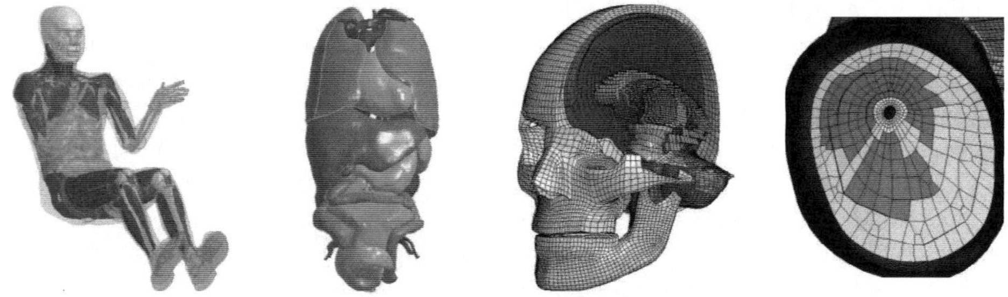

Figure 1. Images of the assembled F05 finite element model

Translational Impact: This study describes the application of a multi-modality image dataset to the development of an anatomically representative small female FEM. The F05-O is the third in the GHBMC family of detailed occupant models (M50, M95) [2, 3]. As such, the F05-O will not only be used for predicting injury risk for small women, but will also provide a tool for evaluating differences in biomechanical response for occupants of varying body habitus. Furthermore, quantitative comparison between various approaches of model output scaling will provide needed insight, particularly as human body modeling is extended to body sizes beyond the M50.

Disclosure Statement: Dr. Gayzik is a Member of Elemance, LLC, which provides academic and commercial licenses of the GHBMC-owned human body computer models.

Acknowledgements: Funding for this project was provided by the Global Human Body Models Consortium, LLC through Project Number WFU-005.

References: [1] Davis et al. Stapp Car Crash J. 60(2016) 509-44. [2] Gayzik et al. ABME 39(2011), 2568. [3] Vavalle et al. Stapp Car Crash J. 58(2014) 361.

Stent Mechanobiology: Impact of Endothelial Shear and Intramural Stress

Henry Y. Chen [a,d], Brian Bigelow [b], Deepak L. Bhatt [c], and Ghassan S. Kassab [a,d]

[a]*California Medical Innovations Institute, Inc., San Diego, CA;*
[b]*Heart Center, St. Vincent Hospital, Indianapolis, IN*
[c]*Brigham and Women's Hospital, and Harvard Medical School, Boston, MA*
[d]*Department of Biomedical Engineering, IUPUI, Indianapolis, IN*

Introduction: Stent can cause flow disturbances on the endothelium and compliance mismatch and increased stress on the vessel wall. These effects can cause low wall shear stress (WSS), high wall shear stress gradient (WSSG), oscillatory shear index (OSI), and circumferential wall stress (CWS), which may promote neointimal hyperplasia (IH). The hypothesis is that stent-induced abnormal fluid and solid mechanics contribute to IH.

Materials and Methods: To vary the range of WSS, WSSG, OSI, and CWS, we intentionally mismatched the size of stents to that of the vessel lumen. Stents were implanted in coronary arteries of 10 swine. Intravascular ultrasound (IVUS) was used to size the coronary arteries and stents. After 4 weeks of stent implantation, IVUS was performed again to determine the extent of IH. In conjunction, computational models of actual stents, the artery, and non-Newtonian blood were created in a computer simulation to yield the distribution of WSS, WSSG, OSI, and CWS in the stented vessel wall.

Results and Discussion: An inverse relation between WSS and IH was found based on a linear regression analysis. A linear relation between CWS and IH was also found. More statistically significant linear relation between the ratio of CWS to WSS (CWS/WSS) and IH was observed ($P < 0.01$), suggesting that both fluid and solid mechanics influence the extent of IH.

Translational Impact: Stents create endothelial flow disturbances and intramural wall stress concentrations, which correlate with the extent of IH formation, and these effects were exaggerated with mismatch of stent/vessel size. These findings reveal the importance of reliable vessel and stent sizing to improve the mechanics on the vessel wall and minimize IH. Translational impact also include device design optimization and surgical planning.

Disclosure Statement: None.

Applicability of Model-Based Design Quality Metrics to Medical Device Software

Paul Jones, Office of Science and Engineering Laboratory, CDRH, FDA, USA
David Hoadley, Ph.D., The MathWorks, Inc., Novi, MI, USA

Introduction: The FDA is faced with the challenge of how a regulator or "third party" (e.g. notified body) can assess the quality, integrity, and ultimately safety and effectiveness of increasingly complex (safety critical) software systems used in these medical devices - under increasingly tight mandated time frames. Establishing the means for a reviewer to quickly establish confidence that a device is likely to perform as intended is a key challenge. We briefly explore technology behind Model-Based Design to see how it might enable reviewer confidence in the behavior and performance of medical device software.

Materials and Methods: To assess the software implementation of medical device features, a manufacturer submits software development life cycle documents/artifacts to FDA for review prior to being placed on the market. This is often voluminous, seldom uniform in organization and content, reflects many person-years of design insight the reviewer does not have, and often provides little visibility into the actual device software. Confidence in device performance is largely derived from an audit of the development artifacts submitted.

Each of the IEC 62304 (the medical device software development life cycle standard) process steps, including associated verification activities (e.g. requirements traceability, consistency, etc.), is very human intensive in practice and fraught with opportunities for introducing errors – even with the aid of quality development process tools (e.g. configuration management, document management, etc.). Evidence of software quality is typically based on testing, which too often provides little insight into design rigor, coding errors, or coverage. Modern techniques such as system- and component-level modeling and simulation of devices before they are implemented in software can reduce the costs associated with creating and verifying a device.

	Software Development Process	Modeling and Simulation	Metrics
IEC 62304, Chapter 5	5.2 Requirements Analysis	Requirement analysis via modeling and simulation	Traceability of simulation test cases to functional requirements
	5.3 Architectural design	Develop structural software models	Allocation of requirements to model components
	5.4 Detailed design	Elaborate architectural model with behavior and software specifications	• Model coverage during simulation testing • Demonstrated compliance to modeling standards
	5.5 Unit Implementation and Verification	• Code generator • Unit models and Software-in-the-loop (SIL) and Hardware-in-the-loop (HIL) testing	• Code ⇔ model ⇔requirements traceability • Unit model vs code behavior (back to back testing) • Structural code coverage • Coding standards compliance • Absence of run-time errors (static analysis) within units
	5.6 Integration and Integration Testing	Subsystem models and SIL testing	• Subsystem simulation vs code behavior • Absence of run-time errors at code interfaces
	5.7 System Testing	System model and SIL testing	• System simulation vs. code behavior • Absence of run-time errors at OS and driver-level boundaries
	5.8 Release	Model configuration management	• Issue tracking/resolution • Change management process conformance

Figure 1: Summary of Model-Based Design stages mapped to the IEC 62304 processes with metrics that can be automatically captured as reports during the verification process highlighted in dark grey.

Modeling and simulation can provide unique and valuable insight into system behavior. The term 'model' refers not only to the abstract representation of patient, physiology, or population behavior (environment), but also the hardware and software to be implemented as the device. In addition to production code, such tools can generate evidence for the implementation steps to demonstrate compliance to a rigorous development process. By automating the artifact creation, one can focus more time on the requirement and behavior analysis to improve the product quality.

Results and Discussions: Here we present Model-Based Design as a rigorous process with metrics that can serve to inform regulators or third parties on the quality, integrity, and ultimately safety and effectiveness, of software systems used in medical devices. Such metrics can form the basis for regulatory science, changing the discussion from an arbitrary quality process level to a concrete product quality level.

Computer-controlled Design and Manufacture of Smart Materials Emulating the Body's Own

Melissa L. Knothe Tate[1,2], Joanna L. Ng[1], Renee Whan[1], Lillian E. Knothe[1], Ulf Knothe[2]
[1]University of New South Wales; [2]TissuTex Party Ltd.

Introduction: Natural tissues such as the periosteum, the hyperelastic sheath covering all nonarticular surfaces of bone, exhibit smart properties[1-3]; *i.e.* they make bone stronger under high impact loads but resilient under lower impact loads with high cycle numbers. The awe-inspiring natural feat of using a "soft sheath" to confer extra strength to "hard bones" under high impact loads is pure genius, albeit not trivial to emulate or replicate on the lab bench. Here, for the first time to our knowledge, we precisely map the intrinsic weave of elastin and elastin in the periosteal sheath of the sheep femur. We render and scale up these natural patterns to create patterns using weaving algorithms, and thus enabling creation of the world's first physical textile swatches emulating the body's own. The resulting smart materials lend themselves for a variety of implantable medical devices.

Materials and Methods: Precise maps of collagen and elastin within periosteum were created in 3D using high resolution multiphoton imaging to excite and capture emission wavelengths intrinsic to collagen and elastin[1] (Fig. 1a,b). These 3D patterns could then be rendered virtually and reassembled using Mimics software (Materialise, MIS v18.0 Beta). Segmentation was carried out, masking each structural protein according to its signal intensity. By converting masks to STL files, textiles emulating periosteum were rapid prototyped representing a composite 3D model comprising collagen, elastin and vascular components (**Fig. 3**). These datasets provided pattern templates for a custom-configured jacquard-weaving algorithm (ArahWeave, arahne CAD/CAM for weaving) and for weaving of physical, scaled up prototypes (AVL Looms, Inc.).

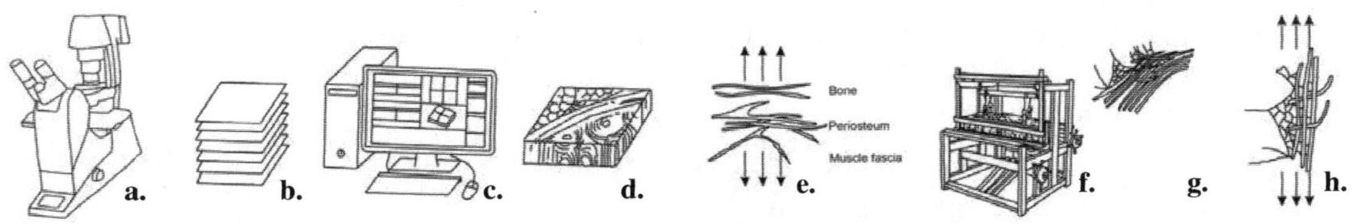

Figure 1. Conceptual work flow for computer-aided design, scale up and manufacture of multifunctional emulating the smart properties of human tissues. (a) High resolution imaging of natural elastin and collagen weave intrinsic to tissue results in (b) high resolution xyz image stacks depicting the precise weave of the structural proteins in the tissue of interest. This intrinsic weave is (c) rendered virtually and (d) reassembled as a virtual tissue using Mimics. Through (e) segmentation, post processing and virtual testing, weaving algorithms are created and optimized. These algorithms can be (f) scaled up and rendered physically using a variety of threads with desired mechanical and other biophysical properties, using a computer-controlled jacquard loom. Resulting multifunctional textiles are then tested and optimized prior to use in next generation surgical implants.

Results and Discussion: A series of textile swatch prototypes were woven using specific combinations of threads selected to mimic collagen (silk and nylon sutures) and elastin (elastane suture). Textiles thus created exhibited viscoelasticity as well emergent properties suggestive of periosteum's natural collagen and elastin weave.[1] Studies are ongoing to determine compositions best suited for surgical repair of bone and other tissue defects.

Translational Impact: These translational studies demonstrate the feasibility and promise of a new class of multifunctional materials inspired by the smart properties of periosteum.

Disclosure Statement: Two co-authors (M.L.K.T. and U.K.) are inventors on intellectual property that has been protected (2 families of patents) and commercialized through university and university spin off (TissuTex Party Ltd.) entities. At the time of abstract submission TissuTex Party Ltd. was in a pre-revenue stage and preclinical studies were underway to determine *in vivo* efficacy of novel textile implants to heal critical sized bone defects.

Acknowledgements: National Health Research and Medical Council.

References

1. Ng J. *et al.*., Nature *Scientific Reports*, 2017, 7, 40396.
2. McBride S.H. *et al. Journal of Biomechanics*, 2011, 44, 1954-9.
3. McBride S.H. *el al. Annals of Biomedical Engineering*, 2011, 39, 1570-158.

Effects of Pore Size on 3D Printed Cast's Mechanical Properties

April Krivoniak, Devin Kiska, Luke Centofanti, Erin Ritzer, Won Joo
Engineering Department, Robert Morris University, Pennsylvania

Introduction: The latest casting materials were developed and implemented in the 1970s. With close to 1.5 million lower arm fractures per year in the United States, a newer version of casting could change the medical field drastically [1]. Improvements in 3D printing over the last 10 to 15 years have provided an avenue to change casting methods. Printing time is the main drawback to 3D printed casts and should be minimized for the method to become mainstream. Adequate strength of the cast must be maintained for protection of the fracture. The main objective in this study is to analyze how altering pore size and therefore the volume fraction affects strength, deformation and printing time for 3D printed forearm casts.

Materials and Methods: Several different technologies and software were used in the study to obtain data. A 3D scanner was used to generate a computer model of a human arm. A cast was then generated from the arm. Pores were added to this solid cast, and pore size and offset were manipulated to generate four different cast models. Finite Element Analysis was run on each model to generate stress data. Each cast was placed in a simulation that would resemble a person falling to the ground. The contact condition between the ground and the cast was simulated. Approximate printing times were calculated by a 3D printer in an engineering lab at Robert Morris University. The volume fraction of each cast was also calculated.

Results and Discussion: As pore size increased, the stress at two different locations (one near the palm of the hand and the other near the elbow) on the cast and deformation increased. Pore size did not affect printing times - all pore sizes resulted in a similar increase in printing time relative to the solid cast. All cast models showed stresses that would not cause failures based on the yield stress of the material. All of the data was formatted as ratios in comparison to the cast without pores.

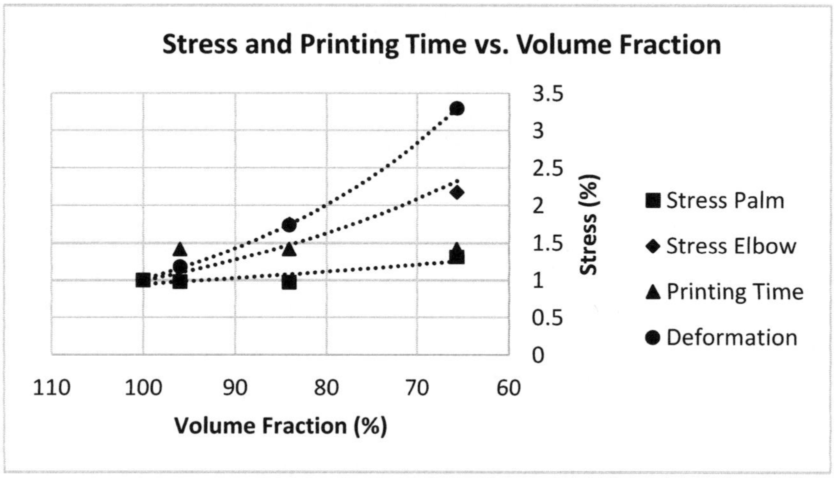

Figure 1. The chart shows trendlines for the stresses at two different locations on the cast models, as well as a trendline for displacement in the Y direction. Printing time ratios are also displayed.

Translational Impact: From the data, it can be concluded that pore size does not affect printing times. Pore size does, however, affect stress and deformation ratios. The larger the pore, the greater the stress and deformation ratios were on the cast models; however, the maximum stress on the largest pore size was still below the yield stress of the material. The data suggests that a smaller pore size would see the least detriment to mechanical properties, but a smaller volume fraction would still be adequate. The simulated results will be verified through mechanical testing being performed on the 3D printed cast with the optimal pore size in comparison with the conventional polyurethane cast.

Disclosure Statement: There are no conflicts of interest within this study.

References:

[1] Journal of Hand Surgery, 2001. *American Society for Surgery of the Hand*, 26A:908-915.

Impact of Modeling Assumptions on Stability Predictions in Reverse Shoulder Arthroplasty

Mehul A. Dharia, Jeff Bischoff
Zimmer Biomet, Warsaw, Indiana, USA

Introduction: Inadequate stability of the baseplate is a leading cause of revision within reverse total shoulder arthroplasty (rTSA). Micromotion between baseplate and bone is commonly used as a pre-clinical indicator for clinical stability (ASTM F2028-14). Finite element analysis (FEA) has been shown to accurately predict baseplate-bone micromotion, but results may be critically dependent on several modeling assumptions. Here, FEA was used to assess the impact of key modeling assumptions related to screw-bone interactions on various rTSA configurations.

Materials and Methods: FEA was used to simulate a fixation experiment. Baseplates of two different sizes (25mm and 28mm diameter), each with a central screw and four peripheral screws, were virtually implanted in a synthetic bone block. Each baseplate was analyzed using 1.5mm and 3.5mm superior-inferior (SI) offsets of the glenosphere center, as well as using four ('4S') and two ('2S') peripheral screws. A clinically relevant loading of 756N was applied in compression as well as in inferior-to-superior shear direction through the glenosphere (Figure 1A). Screw-bone block interactions were modeled in three different ways: (1) Threads were defeatured from the peripheral screws, which were bonded to the bone block (b-nt); (2) Threads were modeled, while still assuming bonded contact (b-t); (3) Threads were modeled, with frictional contact between the threads and bone block (f-t). Micromotion results (Figure 1B) from all 24 simulations (3 screw-bone interactions x 2 baseplate diameters x 2 SI offsets x 2 screw configurations) were compared.

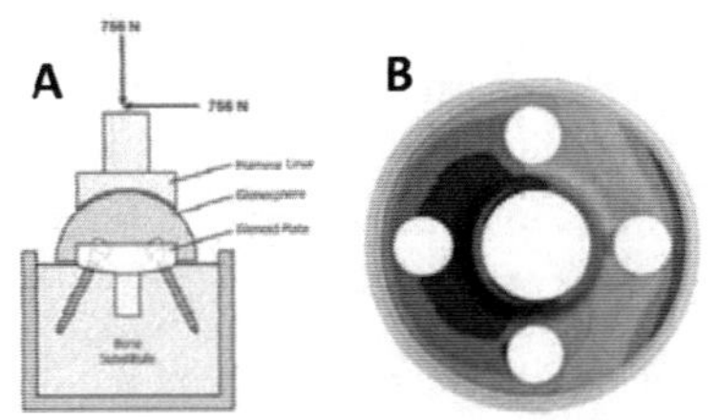

Figure 1. (A) Test Setup (ASTM F2028-14)
(B) Micromotion plot at bone-implant interface

Results and Discussion: Across all 24 configurations, the f-t screw-bone interaction resulted in increased micromotion relative to the corresponding bonded simulations. Screw configuration: For all configurations, the 4S configuration had less micromotion than the corresponding 2S configuration (Figure 2A). SI Offset: For all baseplate diameters, screw configurations, and screw-bone interactions, the 1.5mm SI offset configuration predicted higher micromotion than the corresponding 3.5mm SI offset configuration, except for the 25mm baseplate in b-nt condition (Figure 2B). Baseplate Diameter: For all screw configurations and SI offsets, the f-t modeling assumption resulted in decreased micromotion for the 28mm baseplate as compared to the 25mm baseplate. This trend was reversed for select screw configurations and SI offsets for the other two (b-nt, b-t) modeling assumptions (Figure 2C).

Three different levels of rigor were considered, based on whether or not the screw threads were explicitly modeled, and on the level of friction allowed between the screw and the bone block. Results highlight that answers to basic questions on relative baseplate performance (e.g. is a 25mm or 28mm baseplate more stable?) are sensitive to these assumptions, and require adequate model validation.

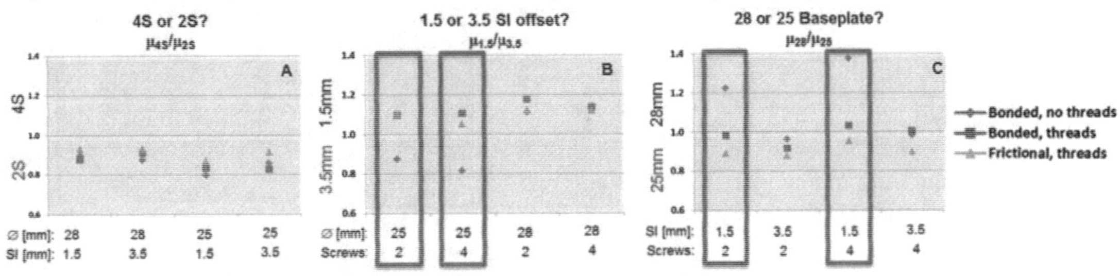

Figure 2. Normalized micromotion comparison among different configurations

Translational Impact: This study highlights the importance of FEA model fidelity (the level of rigor with which the screw-bone interface is modeled) on evaluating differential performance between rTSA baseplate configurations within a single design family. Increased care should be taken when conducting evaluations across multiple device families, when additional variables are present that could confound analyses.

Disclosure Statement: Authors are full-time employees of Zimmer Biomet.

Posing, Morphing and Image Registration of Anatomical Models for MRI RF Safety Assessments

Manuel Murbach[1], Bryn A. Lloyd[1], Alessandro Alaia[1], Esra Neufeld[1], Wolfgang Kainz[2], Fraser Robb [3], and NielsKuster[1,4]
[1]IT'IS Foundation, Zurich, Switzerland, [2]FDA, CDRH, Silver Spring, MD, [3]GE Healthcare, Aurora, OH, [4]ETH Zurich, Switzerland

Introduction: Magnetic resonance imaging (MRI) is a powerful, non-invasive imaging technique. However, the amount of radiofrequency (RF) energy absorbed by the patient – expressed as the specific absorption rate (SAR) – must be carefully managed. The RF exposure is dependent on the patient's placement, anatomy, and posture. Because of the latter two, *in silico* safety assessments are confined by the limited number of computational human models available. Even though there have been major advances in their development, e.g., the Virtual Population (ViP)[1], an increased and parametrized variability in body shapes and postures is indispensable to cover the entire relevant patient population. Therefore, we are developing novel morphing, image registration and posing tools for realistic posture and anatomy adaptation.

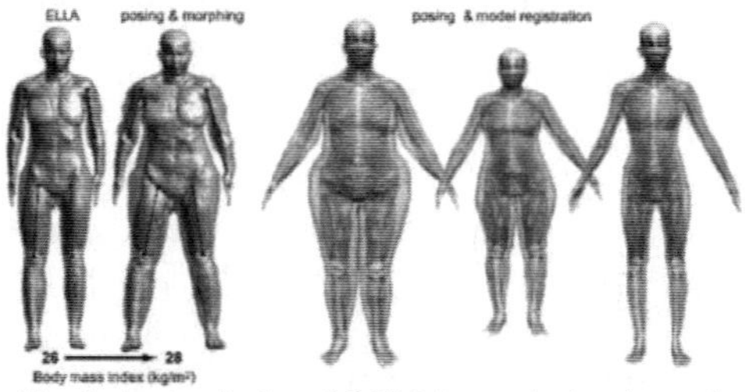

Fig. 1: Anatomical model ELLA morphed and posed from the original shape with a BMI of 26 to a variant with a BMI of 28, and registered onto three different body-surfaces obtained from the CAESAR database [2].

Materials and Methods: Combined morphing and posing of anatomical models is achieved by using constrained biomechanical finite element (FEM) simulations. We have developed a technique for local or regional shrinkage or expansion of tissues, whereby the body is treated as a hyper-elastic material. The method is combined with an approach for changing the posture of the models, whereby bones are moved by prescribed rotations around articulated joints. In both cases, the tissue deformation is constrained by the proximity of nearby rigid bones and regularized by the presence of surrounding soft tissue.

Our approach for 3D registration is suited for known target anatomies, e.g., for personalized assessment. These could include sparse (e.g., cross-sectional) pilot-scan image data of the patient, or body-surface measurements, such as the CAESAR database [2]. This technique relies on non-rigid registration approaches to find a mapping from the computational phantom to patient specific data. Fig. 1 shows examples of all applied techniques, which allow to study the effect of posture and anatomy on the RF and thermal exposure of patients in MRI.

In an initial investigation (Fig. 2), the anatomical model FATS was successively morphed to a similar body mass index (BMI) as DUKE. Additionally, the posture of the arms was adjusted to have comparable distances between the elbows. These evaluations allow – for the first time – to identify in a controlled manner the separate effects of posture and BMI in realistic human models.

Results & Conclusions: The necessary exposure safety margins can be reduced when considering personalized exposure evaluations, allowing for safe extension of the permissible scans for the majority of the patient population. The achieved model-parameterization allows unprecedented insight to the isolated effects of posture and BMI on the RF absorption in MRI, and permits us to better estimate patient-specific RF exposure.

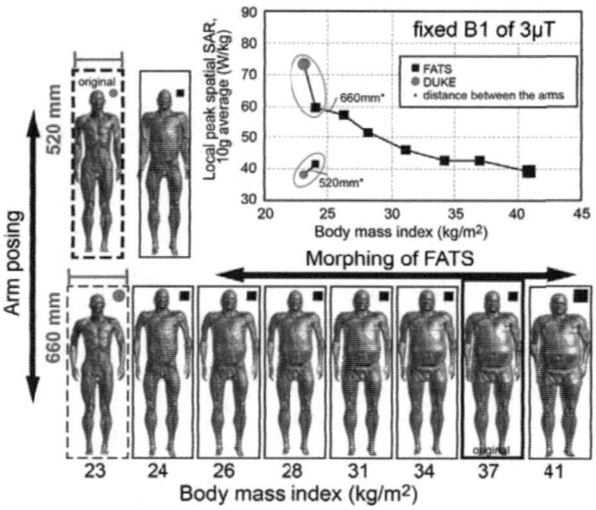

Fig. 2: Local SAR for the different model-variants, for a fixed B1 field of 3μT. Reducing the BMI of FATS results in up to 25% higher local SAR values. The SAR values of DUKE are similar to those of FATS when posed with the same distance between the arms (circled values), indicating that BMI and arm-distance may be important parameters in RF absorption.

Disclaimer: The mention of commercial products, their sources, or their use in connection with material reported herein is not to be construed as either an actual or suggested endorsement of such products by the Department of Health and Human Services. **Acknowledgements:** This work was supported by the MRInext project.

References: [1]. Gosselin M-C. Phys Med Biol 2014;59:5287–5303. [2]. CAESAR. www.sae.org/standardsdev/tsb/cooperative/caesar.htm.

The method of relational correlations for determining subpopulation characteristics

W Andrew Pruett, John L Clemmer, Robert L Hester

Introduction. A new wave of medical therapies seeks to alter chronic physiology by directly or indirectly affecting renal function. These interventions, renal denervation, arteriovenous fistula, and baroreceptor stimulation, alter the balance of a complex network of physiological controls that converge at blood pressure regulation through volume homeostasis. Intra-individual differences in physiology can cause profound variations in the response to these therapies, with the complexity of physiological relationships preventing an easy prediction of responders and nonresponders. Modeling and simulation offer a window into complex physiological systems to see what mechanisms confound or enhance the response to particular therapies. In the integration of mechanisms, synergy can develop to produce or reduce treatment effects, but discerning patterns in high dimensional data requires novel analytic tools. Subpopulations can be identified with binned clustering and the topological data analysis of Singh and Carlsson, but an efficient way to explicitly describe the differences in subpopulations is still lacking. Here, we describe a novel method of describing these differences as linked correlations with multiple ways to visualize and interpret dissimilarities.

Materials and Methods. We use HumMod, an integrative model of human physiology, to create a population of virtual patients who are subjected to a fixed renal denervation (RDN). There a reduction in mean arterial pressure (MAP) of 16$\pm$10.7 mmHg in the virtual cohort, with 30% not responding (showing <10 mmHg fall in pressure). The population was bin clustered, and three different nonresponding populations were extracted. Topological closures were formed over the correlation spaces of each of the nonresponding groups, as well as the entire cohort. These were analyzed for persistent linkages, and the resulting hypothetical mechanism was tested for confirmation in HumMod.

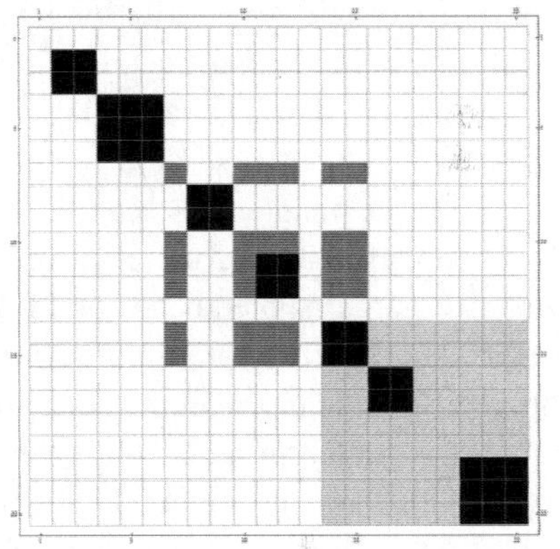

Figure: The correlative structure of Population 1 (dark gray) compared with the whole population (light gray), with overlapping high correlations shown in black. This shows the hormonal and cardiac family (lower right corner) and the urinary, aldosterone, and heart rate family (center).

Results and Discussion. The correlation structure of the full population consisted of four mutually uncorrelated classes: cardiac, blood pressure, renal filtration, and renal excretion. In all nonresponders, angiotensin II was weakly correlated to other variables when compared to the population of responders. In the first nonresponding subpopulation, aldosterone and renal excretion were highly correlated, suggesting a primary role of aldosterone over angiotensin II in blood pressure control, with cardiac preload having a lesser impact on cardiac output. In the second population, extracellular fluid volume and MAP were highly correlated. In the third population, heart rate and atrial pressure were closely linked, suggesting cardiac function was impaired and preload-dependent.

The correlative method, because it allows 22 physiological variables to be accounted for simultaneously, gives a high yield collection of mechanisms for further analysis. This tool completes our program of creating, calibrating, validating, and analyzing *in silico* cohorts for the purpose of clinical trial enrichment.

Translational Impact. This tool identifies physiological mechanisms responsible for enhanced or reduced response to therapy, a critical part of personalizing medicine.

Disclosure Statement. WAP and RLH are employees of HC Simulations, LLC, which owns the exclusive licensing rights for HumMod.

Breast MRI Safety: Simulation of RF Coils with Breast Phantoms Fused to a Human Model

Xin Li[1], Joseph V. Rispoli[1]
[1]Weldon School of Biomedical Engineering, Purdue University, West Lafayette, Indiana, USA

Introduction: Tissue heating caused by high specific absorption rate (SAR) has been a safety concern for high-field MRI breast imaging. However, the SAR distribution among heterogeneous tissue regions, e.g., glandular tissues, chest muscle, fat, etc., is not well understood in prone position breast MRI. To facilitate research and product development complying with US Food and Drug Administration (FDA) guidelines on maximum SAR, we propose a method to simulate in full-wave finite-difference time-domain (FDTD) software the SAR distribution in a heterogeneous breast phantom seamlessly integrated to a human model. We demonstrate results in the 298-MHz RF environment, corresponding to 7 tesla MRI. This method can be used for future study of various types of breast phantoms and body models.

Materials and Methods: Integration of heterogeneous breast phantoms[1] on the pectoral muscle wall of human models is achieved using custom MATLAB functions. First, the pectoral muscle wall is extracted from the torso of the human phantom, and a series of morphological operations are performed on the pectoral muscle wall to smooth the surface. Using affine transformation of 1D axial array of breast phantoms based on the translation parameters from the contours of the pectoral muscle 3D projection, the bilateral breast phantom is seamlessly integrated, creating a whole-body model in the prone position. Physical and electrical tissue properties are assigned based on IT'IS Foundation database. Two previously-reported quadrature breast coils—a combined Helmholtz pair-saddle pair volume coil[2] and a typical half-volume coil composed of two orthogonal and overlapping 19-cm diameter wire loops[3]—are modeled in FDTD software, spaced 1 cm from the nearest tissue. Both coils are driven by 298-MHz sinusoidal current feeds with appropriate 90° phase separation to achieve quadrature operation, and the net input power for each simulation is scaled to 1 W.

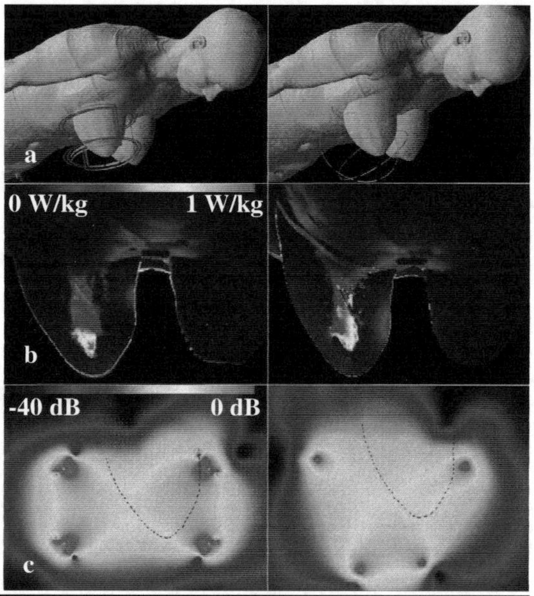

Figure 1. (a) Modeled volume (left) and half-volume (right) breast coils situated around breast phantom fused to whole-body model. (b) Simulated raw SAR, and (c) the rotating magnetic flux density $|B_1^+|$ in the axial plane of a heterogeneous bilateral-breast phantom (0 dB = 7e-05μT).

Results: Figure 1 shows the fused human and bilateral breast model, the two breast coils, the simulated raw SAR, and the rotating magnetic flux density $|B_1^+|$ in the axial plane. The raw SAR plots display localized maximum regions in tissues with high conductivity and relative permittivity, i.e., fibroglandular tissue. The $|B_1^+|$ field shows the distribution of the transmit field in the axial plane of the breast phantoms. The breast phantoms are outlined as dotted lines in the $|B_1^+|$ plots.

Conclusion: As shown in the raw SAR plots, the skin, fibro-connective tissues and pectoral muscle exhibit local maximum SAR. For both simulations, the homogeneity of $|B_1^+|$ inside the breast indicates the transmit field is reasonably uniform. Finally, volume coils tend to create high raw SAR in at both tissue surface and inside tissues, while half volume coil tend to create high raw SAR at tissue surface.

Translational Impact: This simulation demonstrated SAR prediction for two popular breast RF coils. Thus, this methodology has potential to strengthen the MRI safety regulations and compliance for breast imaging.

References: [1] Zastrow E, et al. IEEE T Bio-Med Eng, 2008, 55:2792-2800. [2] McDougall MP, et al. J Magn Reson Imaging, 2014, 40:1165-1173. [3] Bank BL, et al. NMR Biomed, 2013, 26(4), 367-375.

Simulating Disease States with the Living Heart Human Model

Enric Estruch Traveria[1], Jiang Yao[1], Tom Battisti[1], Karl D'Souza[1], Brian Baillargeon[1]
[1]Dassault Systemes SIMULIA Corp, Johnston-RI

Introduction: The Living Heart Human Model (LHHM) is a dynamic four-chamber heart model that considers the interplay and coupling of electrical and mechanical fields acting in concert to regulate the heart filling, ejection, and overall pump functions. The LHHM provides a unique virtual testing environment where medical devices can be deployed and their performance assessed. Because the LHHM represents a healthy heart, ideally it should be modified to represent diseased states in which devices are deployed. The objective of this study is to demonstrate that the LHHM can be re-calibrated to generate physiologically observed responses that are representative of diseased states, such as hypertension, mitral stenosis, and loss of contractility (e.g. due to anoxia).

Material and Methods: The disease states described above primarily impact blood flow and muscle contractility of the heart. The blood flow in the LHHM is modeled using a hybrid approach utilizing both lumped parameter and 3D representations. The 3D model constitutive behavior includes both passive and active responses of the myocardium. In this work, we varied flow resistances to the arterial and pulmonary vasculatures as well as that through the mitral valve. We also varied the ventricular muscle contractility. These variations lead to changes in pressure-volume loops that are representative of physiologically observed relevant hemodynamic conditions of various diseased states.

Results and Discussion: The impact on the Left Ventricle (LV) of increasing the arterial and pulmonary resistances is an increase in the arterial and pulmonary elastance slopes, while its end systolic elastance (Ees) remains constant (**Figure 1**). The simulation also predicts a reduction in the Ejection Fraction (EF) because End Systolic Volume (ESV) is increased. This result is a typical response due to increased afterload (the load that the heart muscle ejects blood against).

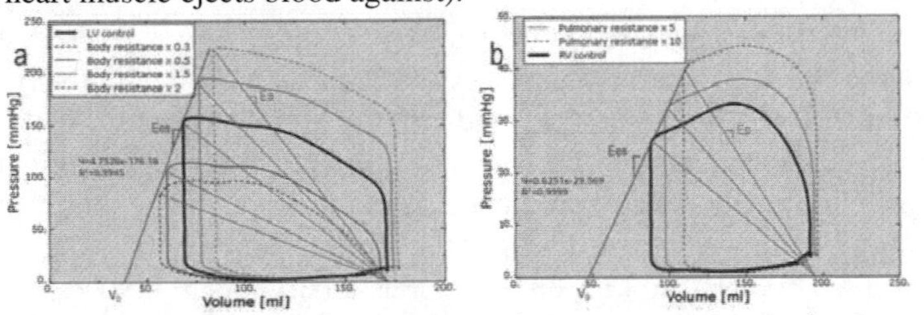

Figure 1: PV loops of (a) LV due to variations of arterial resistance (b) RV due to variations in pulmonary resistance

Increasing the mitral valve resistance (e.g a narrowing mitral valve orifice) impairs LV filling so that there is a decrease in End Diastolic Volume (EDV) or preload. The PV loop still exhibits the same end systolic elastance (Ees) and arterial elastance (Ea) (**Figure 2a**). In this simulation, the afterload was not modified; therefore, the Ea metric is unchanged, but moved in a parallel fashion. However, modifications to the contractility without changes in the afterload or preload lead to different ESVs while nearly maintaining a constant EDV (**Figure 2b**). Therefore, an increase in the contractility is associated with a rise in both the EF and cardiac output while a decrease has the inverse effect.

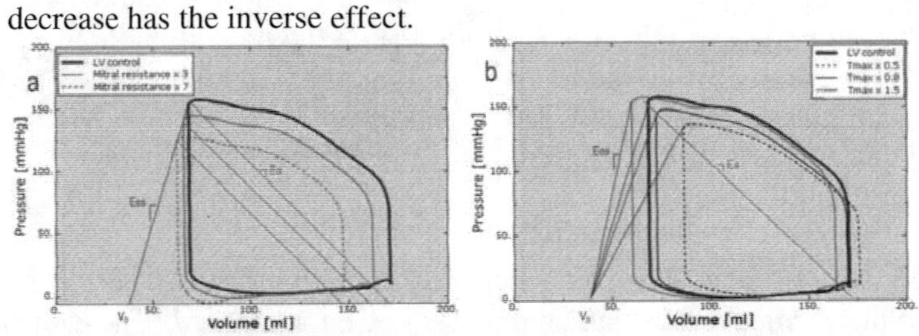

Figure 2: PV loops of LV due to variations in (a) mitral valve resistance (b) muscle contractility

Translation Impact: The results demonstrate the feasibility of utilizing the LHHM to simulate (1) the consequence of arterial and pulmonary resistance changes (e.g. hypertension) (2) the impact of changes in the mitral valve (e.g. mitral valve stenosis) and (3) the impact of decreased muscle contractility (e.g. lack of oxygen, anoxia, or excess CO2, hypercapnia) or increased muscle contractility (e.g. drugs that act as positive inotropic agents), and achieve global hemodynamic responses for the LV that are physiological.

Automatic Construction of Population-Based Models of Left-Ventricle Mechanics

Paolo Di Achille[1], *Viatcheslav Gurev*[1], *Colin B. Compas*[2], *J. Jeremy Rice*[1]
[1]Healthcare and Life Sciences Research, IBM TJ Watson Research Center
[2]Medical Image Analysis for Healthcare Informatics, IBM Research Almaden

Introduction: Despite recent advances in medical imaging, routine clinical exams cannot yet ascertain global kinematics and local strain fields. For example, abnormal contractility and altered hemodynamics cannot be discriminated to guide alternative therapeutic interventions. A correctly informed finite element (FE) model of ventricular mechanics could significantly improve diagnostics. In this study, we propose a novel workflow to automatically build and run high-resolution FE simulations based on geometries extracted from a database of cardiac cine-MR images. Our procedure also addresses two challenges in field to construct FE models from MRI. First, cine-MR datasets have a low resolution that makes their segmentation not directly usable in 3-D simulations. Second the heart is always subjected to loads during the cardiac cycle, while mechanical simulations need to be initialized from the unloaded configuration, which is not measurable from image data.

Material and Methods: An atlas-based segmentation algorithm [1] was used to align and segment wall boundaries from the cine-MR images of the Sunnybrook Cardiac MRI database, a publicly available cine-MRI image dataset [2] (41 patients with 20 images in cycle). Low-resolution segmentations suitable for 3-D simulations are based on an idealized 6-parameter truncated prolate spheroid [3]. A fully automatic procedure was developed to find the optimized geometric models with maximal overlap with the segmented wall boundaries. A meshing algorithm was then used to build discretized computational domains given any 6-parameter set describing left ventricle (LV) geometry. A novel stabilized FE solver of heart mechanics was finally employed to solve for the quasi-static mechanical equilibrium equations governing the contraction cycle [4]. To infer unloaded LV geometries from end-diastolic configurations, we constructed 300 geometries optimally sampled according to the distribution statistics of the analyzed database and pressurize to end-diastolic configurations. A Kriging regression model was then trained on post-inflation geometries as an estimator of unloaded configurations.

Results and Discussion: Figure 1a shows 2 of the extracted parameters over a cardiac cycle that characterize the LV base. Different traces show mean and SEM for the 4 groups of the Sunnybrook dataset (N-healthy subjects, HF-NI-patients with heart failure, HF-I-patients with heart failure and infarct, HYP-patients with LV hypertrophy). Patients with HF-I and HF-NI showed more dilation and thinner ventricles whereas LV thickness was highest in hypertrophy. In Figure 1b, our in-silico methodology predicts that the unloaded state, not directly measurable from patient data, differs substantially from the end-diastolic configuration, where even small pressure cause dilation. Diastolic and beginning of diastole states have traditionally been used as initial conditions of in-silico simulations; our methodology provides a more accurate initial configuration.

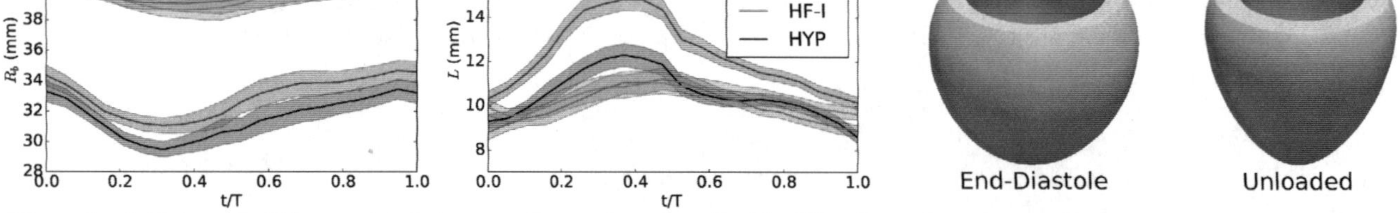

Figure 1. a) Changes in LV geometry during a cardiac cycle as described by LV radius at base (R_b, left), and LV thickness (L, right). b) LV unloading procedure applied to a representative case of a patient with heart failure (HF-I).

Conclusions: We demonstrate a workflow to construct disease-specific mechanics models from databases of cine-MR images and *in silico* simulation to extrapolate to mechanical states that cannot be measured. The workflow can potentially be extended to augment imaging data for improved diagnostics and therapy planning.

References

[1] L. Xie *et al.*, in *Biomedical Imaging (ISBI), 2015 IEEE 12th International Symposium on*, 2015, pp. 376–379.
[2] http://www.cardiacatlas.org/studies/sunnybrook-cardiac-data/
[3] S. F. Pravdin *et al.*, *PLOS ONE*, vol. 9, no. 5, p. e93617, May 2014.
[4] V. Gurev *et al.*, *Biomechanics and Modeling in Mechanobiology*, vol. 14, no. 4, pp. 829–849, Aug. 2015.

Neuro-Functionalized Anatomical Models for Mechanism Investigation, Device Development, and Treatment Personalization

Esra Neufeld[1], Antonino M. Cassarà[1], Wolfgang Kainz[3], Hazael Montanaro[1,2], Niels Kuster[1,2]
[1]IT'IS Foundation for Research on Information Technologies in Society, Zurich, Switzerland; [2]Swiss Federal Institute of Technology (ETHZ), Zurich, Switzerland;
[3]Center for Devices and Radiological Health, Food and Drug Administration (FDA), Silver Spring, USA.

Introduction: Electromagnetic (EM) stimulation is therapeutically applied in electroceuticals/bioelectromagnetic and neuroprosthetic devices that primarily target the peripheral nervous system, as well as for classical neurostimulation of the central nervous system (brain and spinal cord) by implants or externally applied fields. The complex *in vivo* field distribution due to anatomy and tissue variability, as well as its interplay with the highly non-linear neuron electrophysiology and the sophisticated neuromorphology, make it difficult to predict and optimize stimulation regions and treatment effects. Neuro-functionalized anatomical models offer a valuable means to investigate and determine these regions and effects and are thus valuable for investigating stimulation mechanisms, developing and optimizing (novel) therapies and devices, and personalizing therapies.

Materials and Methods: Detailed anatomical models, such as the Virtual Population [Gosselin, et al. Phys. Med. Biol. 59.18 (2014)] and MIDA models [Iacono, et al. PloS one 10.4 (2015)], have been functionalized through integration of electrophysiologically and morphologically realistic cortical pyramidal neurons relevant to transcranial stimulation, selected internal globus pallidus (GPi) and subthalamic nucleus (STN) neurons as well as fibers involved in deep brain stimulation (using DTI fiber tracking), axonal fibers important to spinal cord stimulation (distinguishing dermatomal zones), a vagus nerve (VN) model featuring fascicles, myelinated A- and B-, and unmyelinated C-fibers of importance to VN stimulation, and selected peripheral nerves of concern to low-frequency exposure safety. A novel quasi-static finite element method solver was developed that handles image-based conductivity inhomogeneity and anisotropy information, as well as very thin insulating layers (e.g., epineurium) and is coupled to neuronal dynamics modeling using the NEURON [Hines, Carnevale. Neural computation 9.6 (1997)] library. Functionality was developed to determine stimulation thresholds and recruitment curves, spike initiation locations, as well as exposure-related quantities important for stimulation (activating function along fiber trajectories – predictive of stimulatability; Jacobian Eigenvalue distributions – for worst-case estimation of the region of unintended stimulation).

Results and Discussion: The models and their functionality have allowed i) a novel concept for non-invasive localized brain stimulation via temporal interference stimulation to be discovered, ii) the impact of transcranial stimulation pulse shape on stimulation selectivity and latency (important for learning and long-term effects) to be studied and innovative setups for highly targeted stimulation to be suggested; iii) a novel method to optimize stimulation selectivity with multi-electrode stimulators based on activating function interference matrices to be developed and demonstrated in spinal cord stimulation; iv) assumptions underlying current safety standard to be critically assessed, contributing to the ongoing revision of those standards; v) VN stimulation to be compared with invasive (cuff electrodes), vascular-catheter-based, and external devices with regard to thresholds and recruitment of different fiber types; vi) new insights into deep brain stimulation mechanisms and sites of interaction to be gained, vii) a counter-intuitive major impact of anisotropy on the depth of spinal cord stimulation to be discovered and later explained with an equivalent circuit model; viii) important differences between established axonal dynamics models to be studied, and ix) controversies surrounding time-constant differences observed between electric and magnetic stimulation and their dependence on specific applications to be partially resolved. Verification against reference solution and validation against measurements and literature data on stimulation threshold, localization, and selectivity have been performed.

Conclusions: Neuro-functionalized anatomical models are a valuable tool for mechanism investigation, device development, and treatment personalization. Further application-specific functionalization of anatomical models is required to increase the applicability of coupled EM-neuronal dynamics modeling. Recently begun work on NEUROMAN, a set of high-resolution anatomical human and monkey models extensively functionalized with peripheral nerve models, is expected to be a valuable contribution to the field of neuroprosthetics, electroceuticals, and low-frequency exposure safety.

Acknowledgements: We gratefully acknowledge financial support by the Swiss Commission for Technology and Innovation CTI (S4L-CAPITALIS: 14930.1 PFLS-LS, NEUROMAN: 25290.1 PFLS-LS).

Augmenting a Clinical Study with Virtual Patient Models: FDA and Industry Collaboration on a mock submission sponsored by MDIC

Adam Himes[1], Tarek Haddad[1], Dawn Bardot[2], on behalf of the MDIC mock pre-submission working group participants[3]
Medtronic[1], Medical Device Innovation Consortium (MDIC)[2], Participant list given in: http://tinyurl.com/MDIC-mocksub-group[3]

Introduction: Medical device manufacturers are increasingly using predictive computer models during the product development process, we use the term virtual patient models to refer to these simulations. In some cases, these virtual patient models can be incorporated into a study in a way that is analogous to how some Bayesian clinical trials incorporate historical data as prior information. Benefits of this approach may include increased information from the clinical study, more confidence in the clinical outcome, and in some cases smaller or shorter duration studies.

The Medical Device Innovation Consortium (MDIC) working group on Clinical Trials Informed by Bench and Simulation includes members from industry and the U.S. Food and Drug Administration (FDA). Due to the novelty of incorporating virtual patients into a clinical study, and the mutual interest of both FDA and industry in advancing the use of modeling in the medical device development process, a mock IDE submission was proposed as a means to evaluate and document the process. The sponsor team consisted of members representing medical device manufacturers, engineering software, MDIC, and FDA. A review team was formed at FDA, consisting of members representing engineering, statistical and medical functions, as if the submission were an actual product.

Methods: The working group identified fatigue fracture in a hypothetical new ICD lead (Model 2014VP) as a vehicle for a specific example. ICD lead fracture is a good candidate for virtual patient modeling because it is a failure mode with direct clinical implications and there are examples of analysis methods in the literature [1]. Fatigue performance was modeled by incorporating bench fatigue strength with predicate lead in-vivo use conditions, creating multiple cohorts of simulated virtual patients with fracture survival outcomes. The virtual patients were implemented as prior knowledge in a Bayesian clinical trial design through a novel approach that used a discount function that scales with agreement between real and virtual patients.

The virtual patient model was simulated in a hypothetical clinical study with a failure rate endpoint of 5% at 2 years. We evaluated type I error and power for a range of scenarios. The team followed the standard process for FDA pre-submission meetings to discuss the novel aspects of the procedure, beginning in 2015 and ending in 2016.

Results: Including virtual patients in the clinical trial using the discount function resulted in an increase in both statistical power and type I error. However, due to the discount function, the type I error was much smaller than a for a traditional Bayesian prior, with only moderate decrease in power. The mock submission review process highlighted two key themes: 1) Early communication between the sponsor and review team is important in order to identify areas for detailed discussion, education, or that raise issues of concern, particularly due to the intersection of clinical statistics and engineering modeling. 2) Model credibility and context of use should be integral to the development work, as they will heavily influence the number of virtual patients. The mock submission proceedings are available in [2].

Conclusion: Incorporation of virtual patients into a clinical study can be accomplished by extending traditional Bayesian statistical techniques with a discount function. This approach can provide improved clinical study characteristics compared to no prior information, or without a discount function. The mock submission process provided a mutually beneficial way for the medical device and regulatory community to identify the issues that should be addressed when designing a clinical study that incorporates virtual patients as prior information.

References

[1] T. Haddad, A. Himes and M. Campbell, "Fracture prediction of cardiac lead medical devices using Bayesian networks," Reliability Engineering and System Safety, vol. 123, pp. 145-157, 2014.
[2] http://mdic.org/computer-modeling/virtual-patients/ [Accessed 28 Jan 2017]

Hemodynamics Inside the MonaLSA Stent Graft

R. Tricarico (1), Y. He (2,3), R. Tran-Son-Tay (1,4), S. Scali (2.3), T. Lee (3), S. A. Berceli (1,2,3)
(1) Department of Biomedical Engineering, University of Florida,
(2) North Florida/South Georgia Veterans Health System, University of Florida,
(3) Department of Surgery, University of Florida,
(4) Department of Mechanical and Aerospace Engineering, University of Florida, Gainesville, FL

Introduction: An increasing number of highly sophisticated stent graft for advanced thoracic endovascular aortic repair (TEVAR) are becoming available nowadays. However, highly specialized surgical skills are required for similar thoracic endovascular procedures, and the impact of different stent graft designs on the aortic branch hemodynamics is still under investigation. The objective of this study is to computationally investigate the hemodynamics of various configurations of one of the devices that is approaching the market.

Methods: A post-operatively patient-specific 3D aortic geometry treated with the Valiant MonaLSA device (Medtronic Inc., Dublin, Ireland) was extracted from Computed tomography (CT) scans. *In-silico* modifications were performed to recreate various stent graft designs and placements. Tetrahedral meshes were used for the computational fluid dynamic (CFD) simulations in Ansys Fluent. Stationary and no-slip conditions at the walls were imposed; a pulsatile profile of blood flow, extracted from literature[1] and scaled based on the patient's estimated cardiac output[2], was imposed at the inlet and implemented to obtain the solution of the transient simulations under laminar flow conditions. Patient-specific flow rate waveforms at the branches, and pressure waveform at the descending aorta were imposed as outlet boundary conditions. Intraluminal pressure, time-average and maximum wall shear stress (aveWSS and maxWSS), time-average wall shear stress gradient (aveWSSG), oscillatory shear index (OSI), and relative resident time (RRT) have been analyzed along the aortic branches.

Results and Discussions: A total of nine configurations were obtained through *in-silico* modifications of the design of the stent graft inside the left subclavian artery (LSA), including the *optimal deployment* of the device (**Figure 1, a**). Four configurations were obtained by modifying the length of the stent graft, and four by variating the stent graft curvature; in addition, one configuration with LSA orifice misalignment was tested (**Figure 1**).
Along the stented tract of the LSA, most of the configurations presented both maxWSS (**Figure 3, a**) and aveWSSG values lower than the optimal deployment. The curved designs of 60 degrees, in both forward and backward configurations, showed the lowest values of maxWSS (2.4 ± 1.1 and 3.1 ± 1.4, respectively). All straight designs showed peaks of OSI higher than the optimal deployment. Lastly, the configurations with the highest RRT differences from the *optimal deployment* of the stent graft, had curved designs (*C30B, C60F/B*), while the all straight stent graft designs maintained lower difference ranges (**Figure 3, b**).
On the other hand, on the LSA tract distal to the stent graft, all variables show negligible variation among configurations, except for the RRT, which ranges change depending on stent graft design.

Translational Impact: To the best of our knowledge, this is the first computational modeling of realistic stent graft configurations in a patient-specific aortic arch. Different stent graft designs as well as stent graft suboptimal placement influence the *in-situ* hemodynamics. RRT results to be a useful tool to detect abnormal flow as result of low and oscillatory shear. Analyses are ongoing and WSS-related variables on the aortic wall area around the LSA stent orifice are being investigated.

Acknowledgement: Kind acknowledgments to Medtronic Inc. for granting our research.

References: [1] Gallo, D. et al., *Ann Biomed Eng* 40 (3): 729–41, 2011. [2] DeSimone et al., *Circulation* 95 (7): 1837–43, 1997.

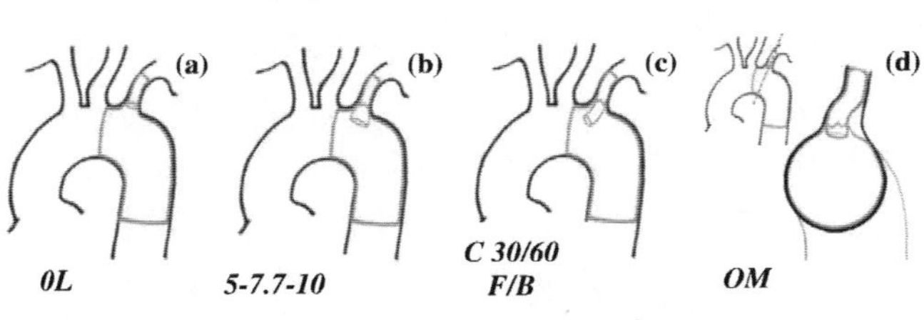

Figure 1. ***In-silico*** **modified configurations obtained in Ansys Spaceclaim (Ansys Inc., Canonsburg, PA) of a MonaLSA device (a,** ***optimal deployment*****).** Down-facing stent graft designs (b, stent graft lengths: 5 mm (*5L*), 7.7 mm (*7.7L*), 10 mm (*10L*)). Forward and backward curved designs (c, lengths: 30 and 60 degree, *C30/60 F/B*). LSA orifice misalignment (d, 30 degrees of rotation to the left, OM).

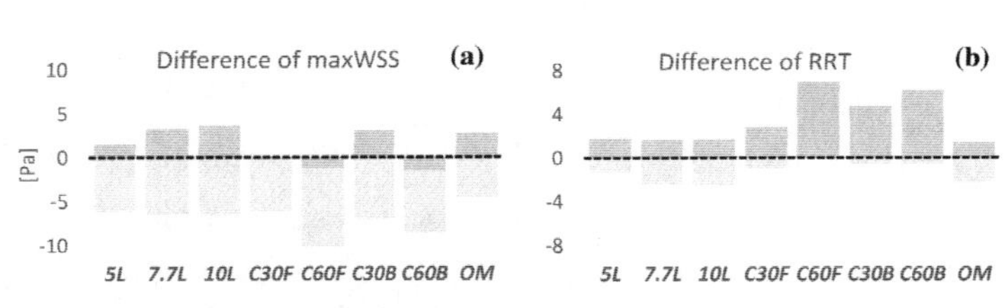

Figure 3. Hemodynamic analysis. Maximum differences between each stent graft design and the *0L* configuration in terms of (a) maxWSS and (b) RRT along the stented LSA. Positive differences indicate larger values on the configuration of interest compared to the *optimal deployment*, vice versa, negative differences refer to larger values on the *optimal* design compared to the configuration of interest.

Biomechanical Effects of Increasing Graft Overlap in Ventral Hernia Repair – Potential Guidelines from a Computational Study

Sahoo S[1], Erdemir A[1,2], Krpata DM[3]

[1]Department of Biomedical Engineering, [2]Computational Biomodeling (CoBi) Core and [3]Department of General Surgery, Cleveland Clinic

Introduction: An estimated 350,000 ventral hernia repairs (VHR) are performed annually in the United States involving a total cost of US $3.2 billion [1]. The use of mesh to reinforce hernia repairs and a mesh overlap of >3-5cm has been recommended to reduce hernia recurrence [2]. However current practices vary widely when it comes to mesh overlap. Currently, there is a paucity of data to inform surgeons about the implications of mesh size on force distribution of the abdominal wall. On the contrary, besides being more expensive, large meshes may cause increased foreign body response, pain and wound morbidity. Thus there is a need to study mesh-tissue interactions to understand favorable amount of mesh overlap and overall mesh size for hernia repair while minimizing the risk of hernia recurrence due to inadequate mechanical fixation. In this study, we aimed to study the effect of mesh overlap amount on abdominal wall force, and tissue and graft stress distributions using a simulation-based approach.

Methods: A 2D finite element model of the abdominal wall hemisection was developed using Abaqus v6.14 (Simulia, Johnston, RI). The model comprised abdominal wall (170mm internal diameter, 15mm thickness), with a midline hernia defect (50mm wide), and a 1mm thick graft fixed under the defect and abdominal wall with 10mm, 25mm, or 50mm overlap. Material properties were adopted from published literature [3,4]. Functional biomechanical outcomes such as midline bulging and midline reaction forces were computed from simulation results. In addition, location and magnitudes of maximum von Mises stresses were also noted in each part.

Results and Discussion: Compared to the normal abdominal wall, the hernia model had increased bulging, higher midline forces and von Mises stresses in abdominal wall (Fig.1). Graft repair mitigated bulging and offloaded midline forces in the abdominal wall/defect, with the graft taking up >2/3 of the load. Moreover, increasing graft overlap lowered von Mises stresses at the graft edge. However, at 5cm overlap, the maximum stress location shifted from graft edge to midline. This observation could be important as it could explain the phenomenon of central graft fractures that are reported with large grafts. Abdominal wall stress was concentrated at the graft edge and, while tissue stresses were lower compared to unrepaired hernia, they did not recover to levels of the normal abdominal wall. Taken together, our results suggest that while greater overlap may reduce stress around the hernia repair, its overall impact on the abdominal wall-mesh construct needs further studies to identify optimal overlap.

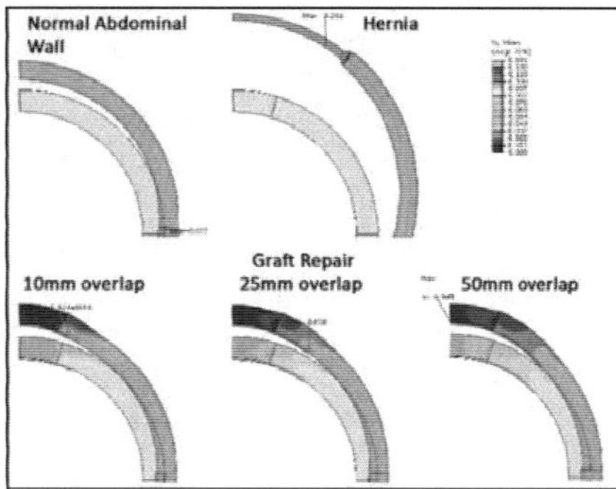

	Bulging	Midline Forces (N/cm)		Max Von Mises Stress (MPa)			
Model	(cm)	Ab wall	Graft	Ab wall	Defect	Graft-edge	Graft-center
Normal abdominal wall	2.2	7.2	-	0.058	-	-	-
Hernia	6.3	8.9	-	0.170	0.201	-	-
Repair-10mm overlap	2.3	1.9	5.5	0.204	0.016	1.979	0.599
Repair-25mm overlap	2.3	2.3	5.1	0.134	0.014	0.636	0.530
Repair-50mm overlap	2.2	2.4	5.0	0.154	0.014	0.538	0.691

Figure 1. Results from models of normal abdominal wall, hernia and repairs with 10mm, 25mm and 50 mm graft overlap.

Translational Impact: This study has the potential to guide surgeons in choosing a more favorable graft overlap to maximize graft performance after hernia repair. The proposed model can inform prospective explorations of graft design and surgical parameters such as graft fixation.

Disclosure Statement: The authors have no conflict of interest.

References: [1] Poulose et al. Hernia 2012, [2] Luijendijk et al. NEJM 2000, [3] Hernández-Gascón et al. Ann Biomed Eng 2013, [4] Pachera et al. J Biomech 2016.

Comparing Modeled Fretting and Observed Wear in Total Hip Arthroplasty Head-Neck Junctions

Andrew P. Baumann[1], Oleg Vesnovsky[1], L.D. Timmie Topoleski[1,2], Finn E. Donaldson[1], Nicole Larsen[1], and Matthew Di Prima[1]
[1]U.S. Food and Drug Administration, Silver Spring, MD 20993
[2]University of Maryland–Baltimore County, Baltimore, MD 21250

Introduction: Total hip arthroplasty (THA) affects over 300,000 people annually in the United States [1]. The *in vivo* fatigue loading of these devices can lead to wear, corrosion, and fretting between mating components [2,3]. This damage can cause adverse local tissue reactions and necessitate revision [4]. Imaging techniques can quantify damage on mating surfaces [5]. Similarly, computational methods can simulate fretting [6]. However, the ability of numerical models to predict damage has not been thoroughly investigated. A validated wear model would allow manufacturers and regulators to predict wear levels without costly experimentation. The goal of this study was to determine the ability of a finite element model to predict wear in THA constructs.

Materials and Methods: Head and neck THA constructs (n = 10) were manufactured out of CoCr alloy (Fig. 1). Construct assemblies were subjected to fatigue loading. Head and neck taper mating surfaces were imaged using the Digital Mosaic Method (DMM) [5]. Construct-specific finite element models were used to calculate the fretting work done (FWD) between the head and neck components [6]. The DMM images and FWD maps were compared. Both domains were re-discretized to have matching resolution and size. Damage in the DMM images was identified with thresholding and discrete Fourier transform (DFT) techniques. The amount and concentration of FWD within damage regions was calculated to assess the model's ability to predict wear in THA constructs.

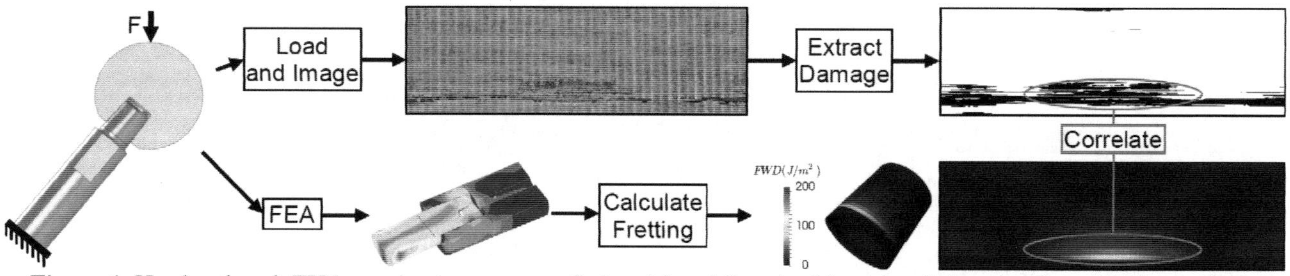

Figure 1. Head and neck THA constructs were manufactured (n = 10) and subjected to fatigue loading. Mating surfaces were imaged and the damage regions were identified. Construct-specific finite element models were used to simulate fretting on the mating surfaces and the results were correlated to the imaged damage regions.

Results and Discussion: Qualitatively, the FWD showed good agreement with DMM images (Fig. 1). Regions of high FWD were within, or in close proximity to, damage regions. Quantitatively, the damage regions contained 12% and 22% of the FWD for the thresholding and DFT techniques, respectively. Additionally, average FWD concentrations within the damage regions were 0.019 and 0.030 mJ/mm^2 for the thresholding and DFT techniques, respectively. This demonstrates the ability of the simulations to predict wear in THA constructs, and illustrates the ability of the DFT approach to identify damage regions within DMM images.

Translational Impact: The results demonstrate that (1) the identification of damage regions within a DMM image can be automated through a DFT approach, and (2) FWD as calculated by finite element analysis can be used to predict wear in total hip arthroplasty head-neck junctions. As such, this approach may be used by device manufacturers to compliment current methods of estimating wear characteristics of THA constructs and to optimize designs for reduced wear. Eventually, after rigorous validation, this approach may benefit regulators by providing a means of evaluating wear potential of total hip devices in regulatory submissions.

Disclosure Statement: The authors have nothing to disclose.

References: [1] Wolford ML, NCHS data brief, 2015, no 186. [2] Goldberg JR, Clin Orthop Relat R, 2002, 401, 149-61. [3] Gilbert JL, J Biomed Mater Res, 1993, 27, 1533-44. [4] Cooper HJ, J Bone Joint Surg, 2012, 94, 1655-61. [5] Di Prima, M, ASTM STP1591, 2015, 147-63. [6] Donaldson FE, J Biomech, 2014, 47, 1634-41.

Using Statistical Interaction Modelling to Predict Cell and Tissue Response to Medical Devices

Danielle Fau, Dr. Brooke Farrugia, Dr. Megan Lord
University of New South Wales

Introduction: The biocompatibility, biostability, and mechanical durability of a foreign material in the body is a key factor to the success of an implanted medical device, and historically it has been difficult to find a material to meet these needs. In the 1990s Dr Thilak Gunatillake and CSIRO colleagues worked to combine the advantageous properties that polyurethanes have to offer (superior mechanical properties and ability to be thermally processed) with the beneficial properties of silicone rubbers (highly biocompatible). These new Elast-Eon™ polymers have since been successfully commercialized and approved by the US FDA for human use (2006). Although in vivo studies have demonstrated the capability of Elast-Eon™ polymers, questions remain on how the polymer performs in vitro in comparison to polyether polyurethanes - a polymer with similar biocompatibility and mechanical properties. A side by side comparison of the Elast-Eon™ polymer and polyether polyurethanes using mechanical testing, and in vitro endothelial and platelet assays will be conducted. And using this data, a model which predicts platelet response to materials will be created by considering the interaction of the following contributing factors: material, sterilization method, and protein coating.

Materials and Methods: Thermoplastic polyurethane elastomer (Elast-Eon™ 2 80A) and polyether polyurethane (Urethane Compound 80A, Pellethane 2363 80A) were solvent cast to obtain flat sheets for testing. FTIR was used to verify the expected chemical structure of the sample materials. Mechanical testing was carried out using a calibrated tensiometer (Instron Mini 5543, MA). Platelet adhesion assays were conducted by exposing the polymer to platelet solution at 37° for 1, 2, or 4 hrs. Platelets were fixed, permeabilised, and stained with rhodamine-phalloidin to stain the F-actin cytoskeleton. Platelets were imaged with fluorescence microscopy and ANOVA analysis was used to determine statistical significance of the platelet count. An endothelial cell (HUVEC) assay was completed using the same method as described for platelets.

Results and Discussion:

FTIR results confirmed the chemical structure of the studied polymers. Mechanical testing demonstrated the difference in mechanical properties between the two classes of polymer (Table 1). Platelets for both polymers had early and late pseudopodial morphology, but polyether polyurethane showed a smaller number of adhered platelets compared to the thermoplastic polyurethane elastomer (Figure 1).

Table 1. Summary table of mechanical properties of polyether polyurethane[1] and thermoplastic polyurethane elastomer

Material	n	E (MPa)	UTS (MPa)
Polyether Polyurethane	2	11.3 ± 0.3	49.5 ± 9.0
Thermoplastic Polyurethane elastomer	6	20.6 ± 3.0	13.8 ± 2.8

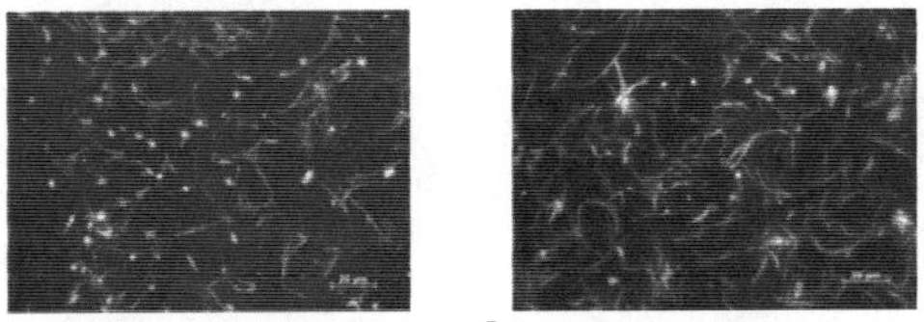

a. **b.**

Figure 1. Fluorescence microscopy images of adhered platelets on Thermoplastic Polyurethane elastomer (a) and Polyether Polyurethane (b)

Translational Impact: These research findings demonstrate the in vitro hemocompatibility of both polymers in investigation and the potential to create a model to predict such interactions.

Disclosure Statement: There are no conflicts of interest associated with this research.

Acknowledgements: Australian Red Cross Blood Service, Whitaker International Program, Australian Research Council Linkage Project Grant, CSIRO, and technical assistance of Zi Yi Zhu.

References:

1. Simmons, A., *Biomat*, 2004, 25, 4887-4900

"Statistical models in optimization of radiotherapy treatments"

Terman Frometa-Castillo[1], Sunyoung Jang[2], Anil Pyakuryal[3], Terence Sio[4], [5]Ernesto Frometa-Leon
[1]Oncology Hospital of Santiago de Cuba, Cuba;[2]Princeton Radiation Oncology;[3]University of District of Columbia;
[4]Mayo Clinic, [5]Faculty of medicine of Matanzas in Cardenas

Introduction: In Radiotherapy (RT) the Treatment Planning Systems (TPS) are medical device. Some of them are biologically based TPS treatment planning system (BBTPS), and use radiobiological (RB) indices: TCP (Tumor control probability) and NTCP (Normal Tissues Complication Probability) for evaluating and/or optimizing treatment in cancer patients.

Additional to the phenomenological/mechanistic TCP/NTCP models, Statistical models project (SMp) has formulated new cell kill and sub-lethal damage (K and SL respectively) formalisms, which involve all possible factors affecting the biological radiation effects (BRE). These formalisms have allowed development of RT-computational simulators aimed to evaluate the TCP/NTCP for RT treatments without using formulas of these indexes, but their own probabilistic definitions [1]

Nowadays Poisson statistics, effective dose (Deff), sigmoid curves, equivalent uniform dose (EUD),TCP Poisson-LQ, NTCP Poisson-LQ and NTCP Lyman are used in three commercially available and most commonly used BBTPS: Monaco® V1.0 (CMS/Elekta, Maryland Heights, MO), Pinnacle® V8.0h (Philips Medical Systems, Andover, MA), and Eclipse V10.0 (Varian Medical Systems, Palo Alto, CA). Many of the RB and probabilistic concepts employed by these BBTPSs described in [2]and RB software [3] represent serious errors and deficiencies in the RT.

The optimization process is based on iterative evaluations, where we find the best treatment aimed to obtain high TCP without violating pre-established complications for the normal tissues (NT) or OARs. The optimization can be done varying the values of imparted dose, number of the fractions or inter-fraction times, which is the methodology used by the software in [3] ; or varying the dose-volume histogram (DVH) such as it is used by the BBTPSs in their IMRT planning inverse modules[2]. Given the NT complication depends on BRE, organ type (serial and parallel) and the regional damage distribution (RDD), a new cost function is proposed for being used in the IMRT module together the values of BRE that represent the mean organ damage, as well as the RDD characterized by BREz and BRE_L, the mean organ superficial and perimeter damage respectively.

Materials and Methods: The three quoted BBTPS were investigated. Macro-probabilistic and RB studies were done for all models and formalisms used by these BBTPS and the recent RT computational simulator evaluating TCP and NTCP.

Results and Discussion: A new cost function is proposed for the IMRT planning inverse modules of the BBTPSs that only uses values related to TCP and NTCP:

$$F(TCP, TCP_0, NTCP_i, NTCP0_i) = \sum \frac{(TCP - TCP_0)}{NTCP_i} * H(TCP - TCP_0) * H(NTCP0_i - NTCP) \quad (1)$$

where TCP0 is the desired TCP specified by the user. Function H is the Heaviside step function. The NTCP0 is limit for the NTCP in each ith associated end-point.

Translational Impact: (a) Development of the methodology for implementing optimization modules into of BBTPS and RT computational simulator; and (b) Determination of errors and deficiencies related to RB and probabilistic concepts in the current RB software and BBTPS.

Disclosure Statement: Conflict of interest: none.

References: [1] Frometa-Castillo T. et al, "The computer-biological RT simulators". Journal of Med Phys, 2016; (under publication process). [2] Report of AAPM TG 166, "The Use and QA of Biologically Related Models for Treatment Planning", March 2012. [3] Uzan J. and Nahum A., "BioSuite. New software for radiobiological customisation of dose and fraction size in external-beam radiotherapy". Presentation at the 10th Biennial ESTRO meeting. Maastricht. The Netherlands, 29 Aug- 3 Sep. 2009.

Ideas Start with People: Fostering the Next Generation of Ideation Engineers

Michele J. Grimm and Deborah Wells
BMES Accreditation Activities Committee

Professional opportunities for biomedical engineers have grown exponentially over the last 10 years and are projected to increase at more than 5 times the rate of other engineering positions between now and 2024 (1). Increases in academic programs in biomedical engineering and bioengineering have followed that trend. Building from the first two programs accredited in 1972, there are currently over 100 accredited undergraduate BME/BE programs located across the U.S. (Figure 1). ABET, Inc. (www.abet.org) is the international accreditation leader for engineering, and the Biomedical Engineering Society (BMES), (www.bmes.org) is the member society for biomedical engineering.

Accredited programs must demonstrate that their graduates meet certain learning outcomes, including the ability to use modern engineering tools. Biomedical engineering curricula must demonstrate that students have the opportunity to analyze, model, design, and realize a biomedical engineering device, system, component, or process while incorporating appropriate engineering standards. These are a sample of the criteria that specifically support the preparation of biomedical engineering graduates for jobs in industry.

Practicing biomedical engineers from academia, government and industry are critical to the accreditation process. Volunteer program evaluators from all three segments of the profession regularly review programs to ensure that accredited programs continue to meet established criteria. Industry and government engineers also play an important role in advising both academic programs and BMES to ensure that program curricula and ABET criteria provide students with the best foundation from which to join the ranks of the biomedical engineering profession. These volunteers help to ensure that the next generation of biomedical engineers are prepared to meet the demands and challenges of the medical device community. Biomedical engineers from industry are today's key to the success of these future engineers. In this presentation, we will tell the audience how they can engage.

Figure 1: Number of Accredited BME Programs

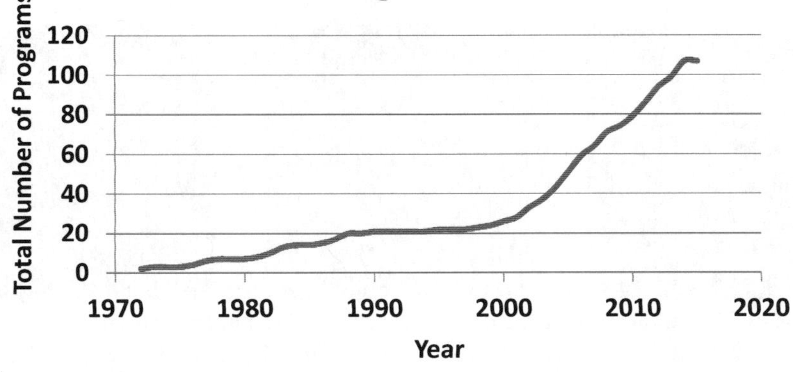

References:

1. Bureau of Labor Statistics, U.S. Department of Labor, *Occupational Outlook Handbook, 2016-17 Edition*, Biomedical Engineers, on the Internet at https://www.bls.gov/ooh/architecture-and-engineering/biomedical-engineers.htm (visited *January 23, 2017*).

A Polyhedral Finite Element Approach to Patient-Specific Modeling and Simulation in Biomechanics

Omar Hafez [1], Mark Rashid [1], Jean-Luc Fattebert [2], Robert Blake [2], David Richards [2]
[1] University of California, Davis, [2] Lawrence Livermore National Laboratory

Introduction: Patient-specific modeling and simulation has quickly become an important analytic and predictive tool in medical applications, ranging from customized and optimized medical device design; improved diagnosis and treatment; surgical design and planning; computer-assisted surgery; and large-scale *in-silico* patient studies. The pipeline of steps in performing patient-specific modeling and simulation are: 1) image acquisition, 2) image processing, 3) image segmentation, 4) mesh generation, 5) simulation, and 6) verification and validation. Polyhedral finite element methods offer exciting improvements in the mesh generation and simulation steps, with promise of pushing the technology further toward a fully automated process. This novel approach is demonstrated in the realm of cardiac mechanics.

Materials and Methods: Medical images are processed and segmented using a combination of thresholding, region-growing, and manual techniques. Watertight surface meshes are extracted from image masks using a novel tolerance-aware Voronoi partitioning scheme. Surface meshes are fed into a robust polyhedral mesh generation tool that handles geometric near-degeneracies for complex and/or evolving domains. A novel polyhedral finite element formulation retains most of the favorable properties of conventional finite element methods, without the same restrictions on element geometry and topology. The approach simultaneously frees the user of mesh quality concerns, generates a higher fidelity mesh of the original geometry, and reduces the system size by up to an order of magnitude. All of which are important improvements when dealing with large problems and/or large datasets.

Results and Discussion: The novel approach is demonstrated for computational cardiac mechanics applications. A polyhedral mesh is generated from MRI data of the human heart ventricles (see Figure 1). The finite deformation constitutive law uses an additive decomposition of incompressible hyperelastic stress response for the myocardium, and an electrical activation-dependent active stress for the muscle fibers. Muscle fiber orientations are generated using a rule-based approach. Fiber activation states are generated from a coupled electrophysiology code. Rigid body modes are removed, and time-dependent ventricular volumes are imposed as boundary conditions. Pressure-volume loops and fiber strains are compared between the results of this work and the Cardioid finite element code for cardiac mechanics.

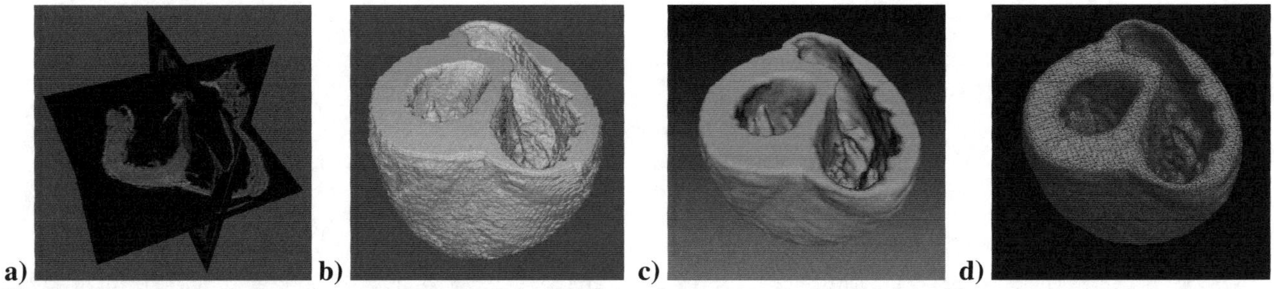

Figure 1. Sequence of steps for generation of polyhedral finite element mesh of ventricles of human heart: a) image acquisition and image processing, b) image segmentation, c) Voronoi-based surface reconstruction, and d) polyhedral mesh generation.

Translational Impact: Polyhedral finite element approaches are showcased as a viable option for computational biomechanics applications as a whole, with significant advantages over conventional techniques. Specifically, the need for manual mesh modification is absolved, and the system size during simulation runtime is significantly reduced. Polyhedral finite element methods have the ability to play in important role in advancing the speed and robustness of image-based modeling in the future.

Acknowledgements: The work of J.-L. Fattebert, R. Blake, and D. F. Richards were performed under the auspices of the U.S. Department of Energy by Lawrence Livermore National Laboratory under Contract DE-AC52-07NA27344.

Doc: A Mobile Training Platform for ECMO and Other Medical Technologies

Dante Archangeli*, John Chirikjian*, Katherine Lin*, Claudia See*, Alyssa Siefert PhD, Tovy Kamine MD, Joseph Zinter III MHS PhD

Yale University. (*these authors contributed equally)

Introduction: Doc is a mobile application and rapid decision-making platform designed to relieve doctors' mental burdens. It reinforces complicated medical technology concepts through guided, real-time troubleshooting. Doctors must sometimes operate a variety of complicated medical machinery, such as extracorporeal membrane oxygenation therapy (ECMO), mechanical ventilation, and other ICU equipment, without adequate practice. To practice, doctors may utilize textbooks, simulators, and workshops, but there are no convenient resources for them to gain adequate experience before encountering potentially patient-life-threatening events. Doc can be a more useful practice tool for physician education. Our prototype focused on ECMO because ECMO complications occur infrequently, so that doctors are often not as experienced in addressing them, but when they arise, they are not only life-threatening, but require quick and effective treatment.

Materials and Methods: To address this medical education gap, we created Doc: an application that defines, troubleshoots, and simulates medical complications by using algorithms similar to those of the digitized "20 Questions" game. Doc's three modes (Search, Diagnose, and Train) enable users to, learn medical terminology, witness the application of knowledge in diagnosis, and practice applying those concepts in hypothetical situations, respectively. Search mode (Figure 1c, 1d) functions as a pocketbook encyclopedia. Diagnose mode asks the user to input an observed symptom (Figure 1a), followed by a series of "yes-no-unsure" questions to identify the underlying complication with its corresponding explanation and treatment (Figure 1b). Train mode presents the user with an adverse, hypothetical patient symptom, from which the user checks for other symptoms (Figure 1e) to identify the underlying complication (Figure 1f). Each mode is a different manipulation of a binary data table, which lists the correlation between a given symptom and complication (1 means the symptom is present for a given complication, and 0 means that it is not). ECMO complications are represented by 30 rows, and potential symptoms are represented by 37 columns. Each complication linked to an explanation and treatment plan. The data table information is collected from a number of peer-reviewed journals and validated by multiple ECMO providers. Doc is built on the Swift programming language for iOS 10+ devices.

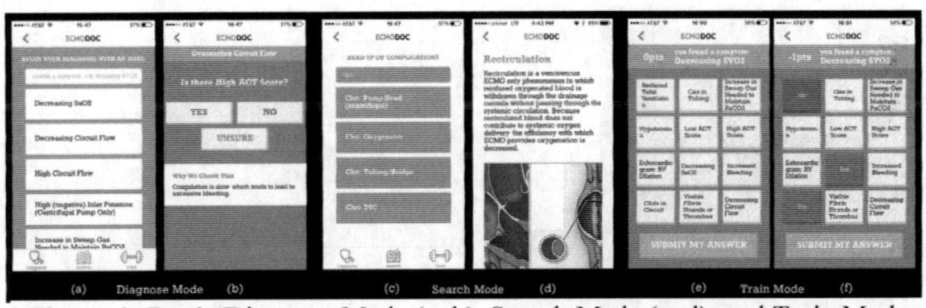

Figure 1: Doc's Diagnose Mode (a, b), Search Mode (c, d), and Train Mode

Results and Discussion: The Doc application was tested with 11 medical students and 4 ECMO providers. Preliminary data demonstrates that Diagnose mode requires on average 5 questions (relating to 37 possible symptoms) to be answered to properly identify one of 30 complications with a confidence of 95%. In Search mode, users were able to access information about a specific ECMO complication within approximately 15 seconds, significantly less time than more traditional search methods (textbooks, journals, web).

Translational Impact: Doc can empower healthcare providers in understaffed hospitals, boost doctor confidence in administering treatments, standardize and improve patient care, and enable more streamlined medical device integration into hospitals. As a promising platform for complex medical technologies and concepts, Doc aims to close gap between the technology and medicine, with numerous quality, economic, and social implications. This platform is translatable to a wide range of concepts and technologies; ECMO is the first step of many. Future iterations will include ICU equipment, mechanical ventilation, Swan numbers and pressors, and more.

Virtual Reality Pathology

Sachith Gullapalli*, Acshi Haggenmiller*, Henry Li*, Alexander Ringlein*, Eben Olson PhD, Richard Torres MD, Maxwell Emerson MS, Alyssa Siefert PhD, Joseph Zinter MHS PhD
Yale University
* Authors contributed equally

Introduction: Limitations of current histology practice include the difficulty of sharing slides, an inability to nondestructively view whole tissue samples, and a restriction to disparate two dimensional (2D) slices of the data. We have developed a novel virtual reality (VR) solution based on data from a volumetric multiphoton microscopy (MPM) technique that circumvents these challenges and provides a real-time interface for viewing and manipulating biopsies in three dimensions (3D).

Materials and Methods: The application was developed using Unity 5.5, a popular game development engine, for use with the *HTC Vive* VR headset and corresponding controllers, as seen in Figure 1. We ran the application on a high performance computer with a dedicated graphics unit (Nvidia GTX 1080) and 32 GB RAM.

Figure 1. demonstration of the use of the *HTC Vive*.

Results and Discussion: Our completed pathology system is able to take a series of two-channel (red-green) images representing slices of a tissue volume imaged via MPM, apply a transfer function to them to mimic a hematoxylin and eosin stain, and render them onto a 3D cube of semi-transparent layers. (Figure 2) The 3D scene is then generated twice, once from the perspective of each eye, and displayed stereoscopically to the user through the VR headset. To interact with the image stack, the system takes user input from the motion controllers and, in a manner similar to multi-touch functionality on touch screens, gives the pathologist the ability to apply slices, translations, rotations, and image transformations to the data. For example, the pathologist may choose to upload a stack of images of a kidney glomerulus, enlarge the cube by dragging it closer to his viewpoint, rotate it to find the best side to work with, slice off the external layers to expose the enclosed glomerulus, and then increase its transparency to visualize the resulting 3D structure.

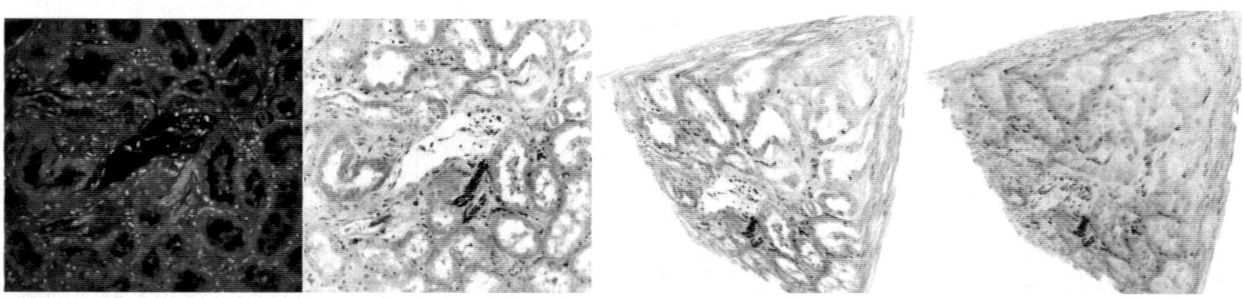

Figure 2. A 2D red-green view (left), with the H&E transfer function applied (second to left), a 3D rendering (second to right), and a 3D rendering with a transparency function applied (right).

Translational Impact: This technology allows pathologists to view renderings of histology data in virtual reality. It provides unique views of the surface and structure of 3D samples which are unachievable using traditional 2D microscopy viewing. It also allows for the viewing of whole, nondestructively imaged samples at the same time. Furthermore, the digitization of whole tissue samples allows for easily sharing samples between pathologists, without the need for physical transportation of tissue. If this application is shown to allow pathologists to be more accurate or timely at making diagnoses, it will likely have a large impact on how the practice of pathology is performed; future studies might compare the effectiveness of our virtual reality technique to more traditional methods of viewing tissue samples.

Disclosure Statement: Richard Torres has an ownership interest in Applikate Technologies, a company that develops 3D tissue imaging.

3D Printed Brain Model for Preoperative Planning and Practice

Peyton Tharp[1], Ryan Branco[2], Savannah Dale[1], Evan Keating[2], Miriam Navarro[1], Jorge I. Rodríguez-Dévora[1,2], Delphine Dean[1]

[1]Department of Bioengineering, Clemson University

[2] Department of Mechanical Engineering, Clemson University

Introduction: Brain scans, such as magnetic resonance images (MRIs) and computer tomography (CTs), provide surgeons with preoperative imagining of a patient's brain [1]. However, during surgery it becomes difficult to visualize the brain matter and structures of interest. Moreover, scanning while performing surgery can be inconvenient and time consuming. Fabrication of a 3D brain model using MRI scans shows potential to increase preoperative planning and practice for neurosurgeons. This project aims to develop a patient-specific, time- and cost-effective, 3D brain printed model using materials that simulates the physical properties of human brain tissue. This will aid surgeons in planning and predicting surgical outcomes with increased accuracy. On this study, we used gelatin and agarose based materials and compared them with porcine brain matter.

Materials and Methods: Using Bose Electroforce 3230 test instrument (load cell 22.2N ±1%), unconfined compression tests were performed on commercial silicone rubbers (from Smooth on), gelatin supplemented with chromium based materials, among brain porcine samples. Samples were 20 mm in diameter and 10 mm in thickness. Porcine samples were dissected from 6-8 month-old pigs weighing 35-40 Kgs. Samples from porcine brains were obtained from anterior, middle, and posterior regions. Concentration of gelatin ranged from 4 to 6 percent supplemented with 1 to 5 grams of chromium. Silicone rubber materials with durometer shore scale 00 from 10 to 20 were also tested. Strain-stress curves were obtained at strain rate of 1/s and maximum stresses at 20% strain were compared.

Results and Discussion: The three regions of the brain elicited significant similar mechanical properties (p-value > 0.05). However, middle region average readings were stiffer than anterior and posterior regions. This can be explained by the most content of white matter compared to the other regions and the fewer folds present in that region. In Figure 1, the strain-stress curves are reported for silicone rubbers and gelatin based hydrogels supplemented with chromium. Stress-strain curves of silicones rubber were excessively stiffer and stronger than the porcine brain matter. The average stress for porcine samples at 20% strain resulted in 1.74 KPa, whereas the closest formulations were 2.17 and 2.53 KPa, for 4%/2gr gel/Cr and 4%/5gr gel/Cr, respectively.

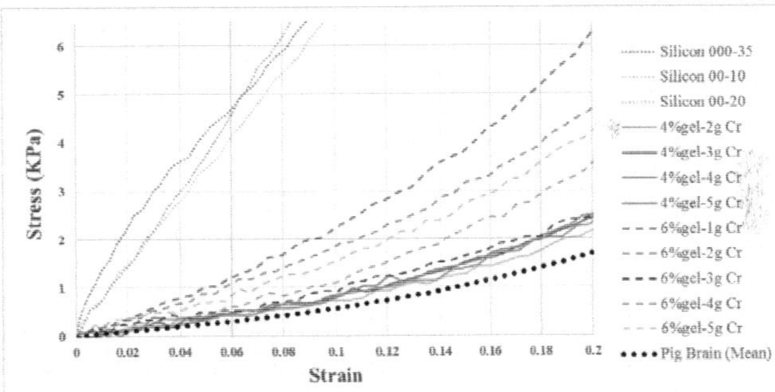

Figure 1. Stress-strain curves for hydrogels (n=5) used to resembled brain mechanical properties. Commercial silicones and gelatin based hydrogel supplemented with chromium. Black dotted line represents the average brain stress-strain curve from 38 brain porcine samples.

Translational Impact: A mechanically relevant brain model promise to provide better simulation tools and experiences to allow neurosurgeon to excel in their practice. Future work involves optimizing the biomaterials used to resemble specific diseases of interest, incorporate vasculature structures and dura mater, and evaluating model efficacy on preoperative planning in a clinical setting.

Disclosure Statement: No conflict of interest are associated with this study.

References: Klein, GT, et al. *World neurosurgery* 80.3 2013.

Acknowledgements: Authors would like to thank the support from the creative inquiry initiative at Clemson University, National Council of Science and Technology from Mexico, and National Institute of Health NIH 5 P20 RR-016461.

SimVascular: An Open-Source Integrated Environment for Anatomic Modeling and Blood Flow Simulation

Nathan M. Wilson[1], Ph.D., M.B.A., Hongzhi Lan[2], Ph.D., Adam R. Updegrove[3], M.S., Shawn C. Shadden[3], Ph.D., Alison L. Marsden[2], Ph.D.
[1]Open Source Medical Software Corporation, Santa Monica, CA
[2]Pediatrics and Bioengineering, Stanford University, Stanford, CA
[3]Mechanical Engineering, University of California – Berkeley, Berkeley, CA

Introduction: Clinical and research interest in imaged-based, patient-specific blood flow simulation has dramatically increased in the last two decades. Three key factors have driven the field: decreasing cost of computer hardware, improvements in the accuracy and resolution of volumetric diagnostic medical imaging, and advances in modeling and simulation from a dedicated and growing field of engineers, scientists, and medical researchers. Now that blood flow simulation has reached the clinic there is more commercial and research interest than ever in the field. While a growing list of commercial packages have added capabilities targeted primarily at device designers and industry, there is a clear need for open-source platforms to enable large-scale research studies to advance academic research. SimVascular, originally released in 2007, was the first integrated open-source environment for realistic image-based blood flow simulation. The NSF funded revitalizing SimVascular beginning in 2013 to eliminate the dependence on three external commercial libraries, modernize the user interface, and provide state-of-the-art modeling tools for the community. To enable large-scale and optimization studies, a generalized vascular model is being developed based on previous work developed in part for the Vascular Model Repository (VMR, www.vascularmodel.org).

Materials and Methods: The SimVascular (www.simvacsular.org) pipeline consists of two primary applications: a front-end for anatomic modeling and an integrated flow solver. The front-end application has recently been overhauled with Qt widgets in part by utilizing code from another open-source effort (MITK, www.mitk.org). The front-end enables users to interact with 3-D imaging data, create anatomic models, perform geometric modeling operations (e.g. virtual surgeries), create meshes, and assign boundary conditions for simulation. An augmented vascular model is then generated using XML with attached data saved using popular standardized and open file formats (VTK, www.vtk.org). A fully integrated open-source blood-flow solver (svSolver, www.svsolver.org) is included within SimVascular. The finite-element based solver includes a custom, scalable linear solver with performance comparable to the previously utilized closed-source commercial library in SimVascular. Linearized wall deformation can be modeled using the Coupled-Momentum Method with spatially varying wall thickness and elastic modulus. Simulation results are typically viewed using the versatile open-source scientific-visualization package ParaView (www.paraview.org).

Results and Discussion: SimVascular and its sister project svSolver are the only fully integrated open-source code packages currently available for image-based patient-specific blood flow simulation. The versatility and broad application of the SimVascular package has been demonstrated with the creation of a public repository, VMR, consisting of over 120 anatomic models and 100 simulations results. Recent improvements in SimVascular include a complete overhaul of the GUI and the elimination of its former dependence on commercial, closed-source libraries. Compilation has been simplified by eliminating Fortran code in the front-end as well as including CMake build support. The full source code is hosted on Github with documentation and samples provided via the websites. New techniques for anatomic modeling in SimVascular include robust techniques for integrating 2-D and 3-D image segmentation enabled by custom robust surface intersection code for triangulated surfaces. Ongoing research efforts include directly creating NURBS patches from discrete triangulated surfaces, improved image segmentation using machine learning techniques, and adding a Python scripting interface to support optimization efforts.

Translational Impact: Since 2013, SimVascular has had over 1,600 unique downloads and has been utilized in over 120 abstracts and journal publications.

Disclosure Statement: No conflicts of interest to disclose.

Acknowledgements: Primary support for this work is provided by NSF SSI Award 1562450.

Integrated Imaging Goggle: Preclinical Studies for Medical Diagnosis and Surgical Navigation

Chris A. Mela[1], Francis Pappay[2], Yang Liu[1]
[1]Department of Biomedical Engineering, The University of Akron, Akron, OH, USA
[2]Dermatology and Plastic Surgery Institute, Cleveland Clinic, Cleveland, OH, USA
Email: cam189@zips.uakron.edu

Introduction: Imaging technologies have been previously developed to help guide medical interventions at the point of care. Often times, these technologies rely on a separate, stand-alone display screen in order to show the image guidance. Such methods can be distracting and confusing, requiring the surgeon to manually correlate the on-screen images with the surgical landscape. To address this issue, we developed a miniature wearable imaging and display system for guiding medical interventions such as surgeries.

Materials and Methods: The Integrated Imaging Goggle utilizes highly sensitive CCD imaging sensors with high efficiency glass lenses to maximize fluorescence detection. The detector unit is affixed to a low-profile head-mounted display unit, Figure 1. We have characterized the detectors resolution, field-of-view, and fluorescent detection limits while controlling the system and recording the real-time video with our custom executable. Additionally, we have integrated real-time ultrasound and pre-operative MRI/CT into the goggle display. We then demonstrated the capabilities of the system through simulated surgical interventions.

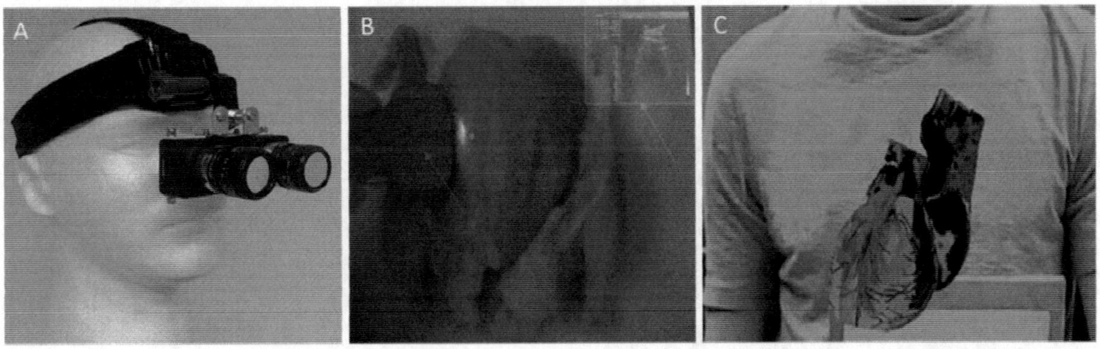

Figure 1 **A.** Prototype Integrated Imaging Goggle. **B.** Goggle image displaying fluorescence (red arrow) with ultrasound (blue arrow). Ultrasound transducer indicated by purple arrow. **C.** Goggle image displaying 3D MRI object registered.

Results and Discussion: The reflectance mode resolution of our detectors was determined to be in the sub-millimeter range. In addition, we have found that the system can detect infrared fluorescent emissions down to a 3nM concentration from typical surgical working distances. We have utilized our prototype system in conducting successful simulated image-guided surgeries. Our system automatically correlates the surgical landscape with medical imaging data in real-time, Figure 1B, C. This information is both recorded by our detector and displayed in the wearable display from the surgeons own line-of-sight, creating a highly intuitive viewing environment.

Translational Impact: The Integrated Imaging Goggle can be an effective tool for diagnosis and guiding surgical procedures. The prototype system is low-profile and highly portable, which holds great potential for various medical situations.

Disclosure Statement: We have no conflicts to declare at this time.

Acknowledgements: This study was supported in part by grants NASA NNX14AL37H, the Ohio Department of Development TVSF 14-503, the LEAP F14-10-LIU-WEB, and The University of Akron Startup Fund.

References: **1.** Mela C, PLoS ONE, 2015, 10, e0141956. **2.** Mela C, Methods in Molecular Biology, 2016, 1444, 85-95.

A Web-Based Platform for Simulating MRI RF-Induced Heating with Implanted Devices

Luca Emili[1], Alessia Baretta[1], Peter Serano[2], Vincenzo Carbone[1], Leonardo M Angelone[2], Marc Horner[3]

[1]Promeditec, Srl., Italy; [2]U.S. FDA, Center for Devices and Radiological Health, USA; [3]ANSYS, Inc., USA

Introduction: Magnetic resonance imaging (MRI) of patients with metal implants (e.g. stents and orthopaedic implants) raises safety concerns because radiofrequency (RF)-induced heating can damage surrounding tissues. Thus, the use of MRI in patients with metallic implants is typically cleared only with specific conditions of use, restricting the accessibility of patients to the benefits of MRI. Computational modeling and simulation (CM&S) use in clinical trials, known as in silico trials, has been limited by the required know-how of simulation experts and costs of computational infrastructure. InSilicoTrials.com is a web-based platform that supports the broadening use of CM&S for electromagnetic safety analysis prior to medical device submission, with the goal of extending the adoption of MRI as a diagnostic tool for patients with implanted devices.

Materials and Methods: A web platform was developed to automate the set up and solution of the ASTM F2182 test method [1]. As shown in Fig. 1, the computational model was comprised of a simplified stent, an ASTM phantom, and a 64 MHz RF transmit coil [2]. The RF field and specific absorption rate (SAR) were calculated using the finite element solver ANSYS HFSS v17.2. The web interface is pre-populated with the required inputs so that users can easily change simulation input parameters, e.g., device material and length, phantom material, coil dimensions, and radio-frequency (Fig.2). Simulations are performed in the cloud and a summary report based on current regulatory guidelines is provided upon in silico trial completion [3]. Simulation results are stored in a secure and reliable environment controlled by the user.

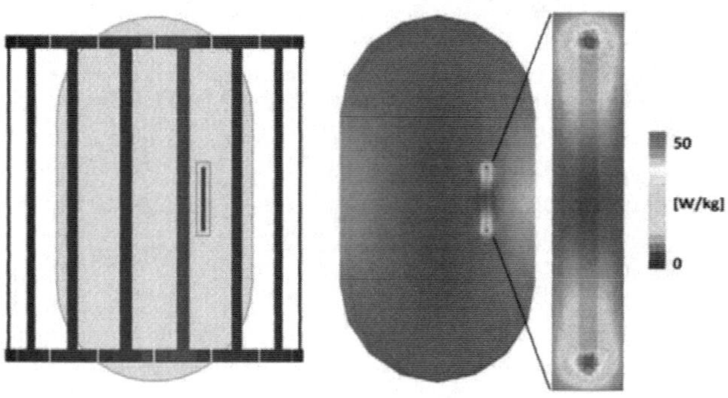

Figure 1. Model geometry (left) and computed SAR (right)

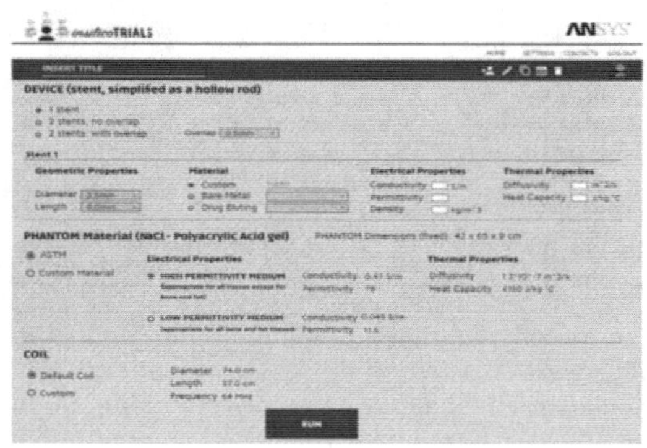

Figure 2. Platform InSilicoTrials.com

Results and Discussion: The main advantage of the in silico trial platform is that it ensures that simulations are executed according to established simulation best practices. Another advantage is its usability, with no specific expertise in numerical methods or numerical solvers required. Moreover, it eliminates the need for extensive hardware infrastructure since all simulations are run remotely on the cloud.

Translation Impact: The proposed platform promotes the broader adoption of simulations in preclinical trials for RF safety analysis, supporting the medical device submission process and pre-market regulatory evaluation [4].

Disclosure Statement: The mention of commercial products, their sources, or their use in connection with material reported herein is not to be construed as either an actual or implied endorsement of such products by the Department of Health and Human Services.

References

[1] ASTM F2182-11a, Standard Test Method for Measurement of Radio Frequency Induced Heating On or Near Passive Implants During Magnetic Resonance Imaging; [2] Serano P et al. BMES/FDA FMD Conference 2015; [3] FDA Guidance: Reporting of Computational Modeling Studies in Medical Device Submissions, 2016; [4] Morrison TM et al. BMES-FDA FMD 2016, Special Issue of Annals of Biomedical Engineering.

Ten "Not So" Simple Rules for Credible Practice of Modeling & Simulation vs Considerations of Medical Device Development Tools Program

Ahmet Erdemir, Gary An, Jeff E. Bischoff, David M. Eckmann, Marc Horner, C. Anthony Hunt, Joy P. Ku, Donna R. Lochner, William W. Lytton, Vasilis Z. Marmarelis, Alison L. Marsden, Tina M. Morrison, Lealem Mulugeta, Jerry G. Myers, Grace C.Y. Peng, Gaurav N. Pradhan, Martin J. Steele, Rajanikanth Vadigepalli, Marlei E. Walton
Committee on Credible Practice of Modeling & Simulation in Healthcare

Introduction: Computational modeling & simulation (M&S) is penetrating biomedical research and the medical device industry at an accelerated pace because M&S offers a platform to design interventions by providing the capacity to effectively evaluate performance and safety. Yet, the promise of M&S for scientific discovery and clinical care can only be realized when modeling and simulation workflows are credible. Recently, a multidisciplinary committee, the Committee on Credible Practice of M&S in Healthcare (CPMS), was formed to provide guidance for quality assurance that spans the spectrum of M&S activities. CPMS released Ten "Not So" Simple Rules (TSR) that recognize various credibility assessment procedures throughout the lifecycle of M&S, and are agnostic to the desired use and medical discipline of M&S [1]. The US Food and Drug Administration (FDA) also acknowledges the need for accessible and dependable models to accelerate the translation of medical innovations to market. In 2013, draft guidance on the processes for the qualification of medical device development tools (MDDT) for use in device design and evaluation was released [2]. One category of MDDT is the nonclinical assessment model, which includes computational models for the simulation of device function. The pressing regulatory science question is the practicality and utility of community perspectives in assuring M&S quality at a regulatory level. Therefore, the goal of this study was to understand the correspondence of TSR with the considerations of MDDT.

Materials and Methods: A comparison was performed between TSR and MDDT (Table 1).

Table 1. Ten "Not So" Simple Rules (TSR), by the Committee on Credible Practice of Modeling & Simulation (M&S) in Healthcare [1] vs. considerations noted by the US FDA for qualification of a medical device development tool (MDDT) [2]. Color coding indicates direct correspondence.

TSR	MDDT
Rule 1 – *Define context clearly.*	Description of MDDT
Rule 2 – *Use appropriate data.*	
Rule 3 – *Evaluate within context.*	Context of use
Rule 4 – *List limitations explicitly.*	
Rule 5 – *Use version control.*	Strength of evidence
Rule 6 – *Document adequately.*	
Rule 7 – *Disseminate broadly.*	Assessment of advantages and disadvantages
Rule 8 – *Get independent reviews.*	
Rule 9 – *Test competing implementations.*	Consent to public disclosure and use
Rule 10 – *Conform to standards.*	

Results and Discussion: There are identifiable overlaps between TSR and MDDT. The documentation efforts recommended by the community (Rule 6) align with the requirement for description of the MDDT (Table 1). Similarly, concerns regarding the context of use are recognized in both efforts (Rule 1 vs "Context of use" in Table 1). Activities to obtain adequate data and to establish the predictive capability of the model (Rules 2-3) can also inform content for strength of evidence for the MDDT (Table 1). Explicitly documenting limitations (Rule 4) can help assess advantages and disadvantages of the model as part of the qualification package (Table 1). An open development approach and public dissemination efforts, which are recognized community-based rules (Rule 7), will further facilitate public disclosure and use of the MDDT upon its qualification (Table 1).

Translational Impact: Identification of correspondence between TSR and MDDT can facilitate preparation of qualification documentation for submission to the US FDA.

Acknowledgments: The authors would like to acknowledge the Interagency Modeling and Analysis Group and the Multiscale Modeling Consortium, who enabled activities of the Committee. Detailed author affiliations can be accessed at https://simtk.org/plugins/moinmoin/cpms/CPMS%20Members.

References: [1] Erdemir, A., Mulugeta, L. and Lytton, W. W. Ten "not so" simple rules for credible practice of modeling and simulation in healthcare: a multidisciplinary committee perspective, 2015 BMES / FDA Frontiers in Medical Devices Conference: Innovations in Modeling and Simulation, May 18-20, 2015, Washington, DC. Available at: https://simtk.org/websvn/wsvn/cpms/doc/posters/fmd_2015.pdf [2] US FDA, CDRH. Medical Device Development Tools - Draft Guidance for Industry, Tool Developers, and FDA Staff. Available at: http://www.fda.gov/MedicalDevices/DeviceRegulationandGuidance/GuidanceDocuments/ucm374427.htm

Fusion and Non-Fusion Spinal Implants: Computational and Experimental Study

Mary Foltz[1-4]; Arin Ellingson[3]; Andrew Freeman[2]; Joan Bechtold[1,2,4]; Victor Barocas[2]; David Polly, Jr[1]
[1]University of Minnesota, Orthopaedic Surgery, [2]Biomedical Engineering, [3]Rehabilitation Medicine, [4]Excelen Center

Introduction: Non-fusion, or growing-rods, are commonly used to accommodate a growing spine for curvature correction due to early onset scoliosis. High failure rates are associated with non-fusion instrumentation; most prevalent (15-20%) in rod failure at rod-connector interactions[1]. The instrumentation uses fewer anchors and spans greater distances while there is increased load shared with construct relative to fusion instrumentation used in mature spines. Hence, a need exists to quantify mechanical performance of non-fusion constructs and compare to fusion constructs. ASTM F1717 is the current standard for testing mechanical performance of fusion constructs – this standard was used for testing fusion constructs and modified for non-fusion constructs. Computational models allow for enhancement, manipulation, optimization, and analysis of pre-submission communication with the regulatory office. The purpose of this study was to experimentally determine structural properties of fusion and non-fusion constructs; use the data to develop and validate finite element analysis (FEA) models.

Materials and Methods: Fusion and non-fusion (growing-rods with side-by-side connectors) constructs were mechanically tested per ASTM F1717 (n = 5/group) using pedicle screws (TI-6Al-4V & Co-28Cr-6Mo), set screws (Ti-6Al-4V), spinal rods (Ti-6Al-4V), test blocks (UHMWPE), and connectors (Ti-6Al-4V, non-fusion only). Blocks for non-fusion were offset to incorporate connectors, placed at the mid-point between the superior and inferior blocks. Experimental test methods were simulated with FEA by ABAQUS; material properties were obtained from ASTM standards (Ti-6Al-4V & Co-28Cr-6Mo) or through MakeItFrom material properties database (UHMWPE) (Figure 1A). FEA assumed isotropic and linear-elastic for blocks, while the instrumentation was assumed to be isotropic and linear elastic-plastic. Perfect surface-to-surface contact was assumed for all surface interactions. Stiffness, yield strength, and ultimate strength were measured experimentally and through FEA. Fusion and non-fusion constructs were compared by the Student's t-test; FEA was validated when within 10% difference from the experimental.

Results and Discussion: FEA predicted stiffness and ultimate strength of both fusion and non-fusion constructs within 10% of experimental (Figure 1B, Table 1). However, the yield region deviated from experimental. Non-fusion constructs had a significantly greater ultimate strength and stiffness than fusion (Table 1). Stress distribution analysis illustrates concentrations at apex of curvature for fusion construct and at superior rod-connector interaction for non-fusion.

Table 1. Mechanical Characteristics – [1]stiffness, [2]yield strength, [3]ultimate strength, [4]non-fusion, [5]fusion, and [6]experimental († and ‡ represents significance at $p<0.01$)

	S[1], N/MM	YS[2], N	US[3], N
NF[4], EXP.[6]	$48\pm2^{\dagger}$	475 ± 60	$1080\pm9^{\ddagger}$
F[5], EXP.[6]	$43\pm0.3^{\dagger}$	457 ± 17	$1007\pm6^{\ddagger}$
NF[4], FEA	47	541	1035
F[5], FEA	44	575	990

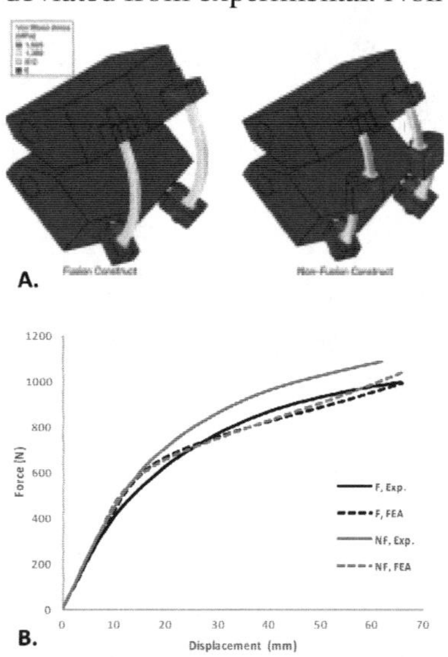

Figure 1. (A) FEA Fusion and Non-Fusion; (B) Comparison of Experimental and FEA for Fusion and Non-Fusion

Translational Impact: Structural properties agreed for stiffness and ultimate strength between experimental and FEA for both constructs. Though these constructs distribute stress differently, their yield strengths were not significantly different. Stress concentrations observed on the non-fusion computational model were consistent with clinical observed failures in the literature, where rod breakage occurs within 1cm of the rod-connector interaction[1]. Similarities observed between clinical failures and FEA stress concentrations provides a basis for utilizing computation modeling in regulatory submission. Future work is to enhanced understanding of material properties, particularly the yield properties.

Acknowledgements: Thanks to Medtronic, Inc. for donating hardware and the Minnesota Supercomputing Institute. Support provided by NIH/NICHD (K12HD073945).

References: [1] Bess S et al., *J Bone Jt Surg.* **92**(15):2533-43.

Pre-clinical Testing of 3D Printed, Nanostructured Osteochondral Implant for Knee Repair in a Small Animal Model

Ben Holmes[1 3] bbh2c@gwmail.gwu.edu, Che-Ying Kuo[2 3] cykuo238@umd.edu, Navein Arumugasaamy[2 3] navein@umd.edu, Nidhi Gandhi[2 3] ngandhi04@gmail.com, Christopher Rossi[3] ctrossi@childrensnational.org, Johnathan Fisher[2 3] jpfisher@umd.edu, Lijie Zhang[4] lgzhang@gwu.edu, Matthew Oetgen[3] moetgen@childrensnational.org

1. Nanochon LLC, Burke VA, 2. University of Maryland, College Park MD, 3. Children's National Medical Center, Washington DC, 4. The George Washington University, Washington DC

Introduction: Osteochondral lesions of the knee are difficult injuries to treat. Despite improvements in the diagnosis of these lesions, optimal treatment remains elusive, likely as a result of the complex interactions between host factors and lesions specific factors. Lesions with disrupted cartilage that are unstable are especially difficult to treat in skeletally immature patients with current treatment methods leading to mixed results overtime. We investigated the feasibility of using three-dimensional biologically inspired implants, manufactured using novel 3D printing techniques and synthetic bio-nanomaterials for treatment of osteochondral defects in a rodent model.

Materials and Methods: An osteochondral defect was created in the trochlear groove of the left knee of 6-week old female Sprague-Dawley rats. 2 mm defects were created. Four experimental groups included a blank control, a solid hydrogel implant and two experimental implants were implanted. Two different experimental implants were designed mimicking the complex micro-structure of the osteochondral transition zone and fabricated using a 3-d bioplotter: 1) a polycaprolactone and polyethylene glycol blended resin which is photocured via UV light and a nonoporous thermoplastic polyurethane (TPU) based material. At 1 and 3 months samples were stained for hematoxylin and eosin (Figure 1).

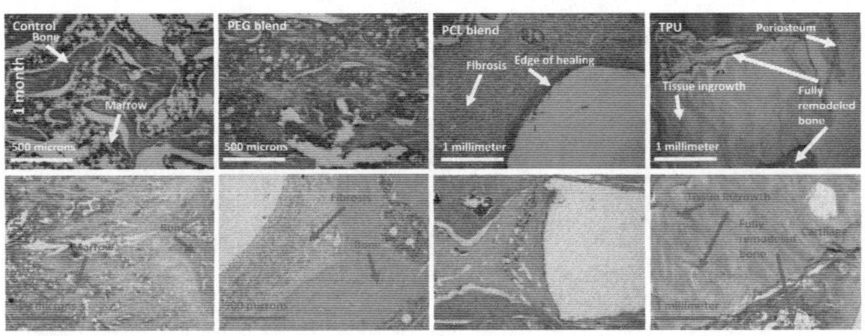

Figure 1. Histology of tissue samples harvested after 1 month(top) and 3 months (bottom), comparing a blank defect, polyethylene glycol (PEG) hydrogel implant, polycaprolactone (PCL)-hydrogel composite and a 3D printed nanoporous thermoplastic polyurethane (TPU) implant.

Results and Discussion: The surgical procedure was successful in all animals and all subjects survived to end point. Histologic analysis (Figure 1) demonstrated some bone remolding and mostly formation of trabecular bone and marrow in the control, and in the PEG group. The PCL based experimental implant showed the formation of extensive fibrosis and scar tissue. TPU showed the formation of fully remodeled bone around the implant, and some bone invading the implant microstructure. As periosteum like membrane also formed over the articulate surface of the implant. At 3 months this yielded the formation of new fibrocartilage over the implant surface, and fusing the implant to existing articulate cartilage. Additional *in vitro* study was performed to evaluate vascularized bone formation.

Translational Impact: Experimental highly nanoporous TPU material can effectively be 3D printed into a complex microstructure implant for the repair of osteochondral lesions, accelerating bone growth with no fibrosis, as compared to controls and other experimental designs. The implant also showed signs of encouraging vascular and cartilage formation. There is great potential for a scaled up implant which could quickly graft to bone and fuse to the healthy cartilage in a human patient's defect, for fast and effective recovery from joint damage/disease.

Disclosure Statement: Work presented in this study may directly impact the value of intellectual property and holdings in Nanochon LLC, which would directly benefit some of the authors.

Acknowledgements: The authors would like the thank The Bioengineering Lab for Tissue Engineering and Nanomedicine at George Washington University, The Sheikh Zayed Institute for Pediatric Surgical Innovation and The Department of Orthopedic Spots Medicine at Children's National Medical Center and the Pediatric Orthopedic Society of North America for their talents, support and funding.

Minimally Invasive Delivery and Imaging of Therapeutic Stem Cells in the Lung

Jinho Kim[1], John D. O'Neill[1], Brandon Guenthart[2], Matthew Bacchetta[2], and Gordana Vunjak-Novakovic[1,3]
Departments of [1]Biomedical Engineering, [2]Surgery, and [3]Medicine,
Columbia University, New York, NY 10027, USA

Introduction: Approximately 10 million people worldwide die each year from lung disease, including lung infection, lung cancer, and chronic obstructive pulmonary disease, accounting for roughly one-sixth of all deaths globally [1]. Stem cell-based lung treatments have emerged as promising therapeutics that can significantly impact clinical outcomes for patients with serious lung conditions [2]. However, only marginal efficacies of stem cells have been reported in human clinical trials, largely because of an incomplete understanding in the transport mechanisms of cells following delivery, for example, into the lungs and a lack of effective ways to monitor the fate of the cells (e.g., engraftment, proliferation) within the lung. Here, we present our studies on the transport dynamics of stem cells administered intra-tracheally via liquid instillation, and a minimally invasive approach for real time *in situ* imaging of cells within the lungs using optical fiber imaging probes.

Materials and Methods: Mesenchymal stem cells (MSCs) derived from the bone marrow of Sprague-Dawley rats were used. The cells were fluorescently labeled with carboxylated Qdots (Q21321MP, Thermo Fisher Scientific) or carboxyfluorescein succinimidyl ester (CFSE, ab113853, Abcam) and suspended in culture medium (concentration: $\sim 2.0 \times 10^7$ cell/mL). A fiber-optic imaging probe (FIGH-03-215S, Fujikura) integrated into a custom-built fluorescent microscope was employed to visualize the luminal airway surface. A micro-prism mirror (8531-601-1, Precision Optics) was attached to the distal end of the probes for visual inspection of airways.

Results and Discussion: We first investigated mathematically the transport and deposition mechanisms of cells administered into the lungs through the airways (**Fig. 1a**). During liquid instillation in clinically relevant situations, MSCs will readily follow liquid flow because cell Reynolds number $Re_C \sim 10^{-4}$, Stokes number $Stk \sim 10^{-5}$, and $F_V/(F_G - F_B) \sim 10^3$ where F_V, F_G, F_B are, respectively, viscous force, gravity, and buoyancy. An injected liquid can deposit a liquid film with thickness $h \sim Ca^{2/3}$ on the surface, where Ca is the liquid capillary number, resulting in increased cell seeding density with Ca. Fiber-optic imaging probes were prepared to visualize the interior of the rat trachea following instillation of Qdot-labeled MSCs through the airways (**Fig. 1b**). By inserting and translating the imaging probe distally within the trachea (**Fig. 1c**), the carina, and left and right bronchi were clearly visible (**Fig. 1d**). In addition, an even deposition of the Qdot-labeled MSCs on the trachea surface was visually confirmed using the imaging probes (**Fig. 1e**). Histologic images of the distal lung showed that CFSE-labeled MSCs were attached on alveolar surfaces (**Fig. 1f**).

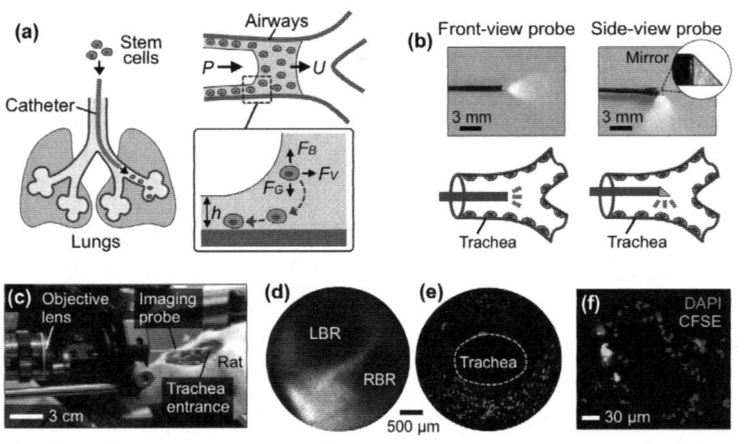

Figure 1. Stem cell delivery and imaging in the rat lungs: (**a**) Mechanism of cell transport and deposition. (**b**) Imaging probes used for imaging airways. (**c**) *In situ* imaging of the rat trachea. (**d**) A bright-field image (LBR: left bronchus, RBR: right bronchus) and (**e**) fluorescent image of the rat airway interior, and (**f**) histology image of the rat alveoli following a MSC administration.

Translational Impact: Our findings and approach for minimally invasive delivery and imaging of mesenchymal stem cells in the lungs may be useful in addressing the burden of lung disease by establishing more effective strategies for stem cell-based treatments for lung diseases and injuries.

Disclosure Statement: The authors declare no conflicts of interest.

Acknowledgements: Financial support from NIH (HL120046, EB002520) and Mikati Foundation

References: [1] T. Ferkol et al., *Ann Am Thorac Soc*, 11, 404-406 (2014); [2] D.J. Weiss et al., *Stem Cells*, 32, 16-25 (2014).

Prospective Design of Combination Products/Delivery Devices Exploiting Multiphysical Mechanisms for Wound Healing

M.L. Knothe Tate, J. Rnjak-Kovacina, P.B. Zetterlund, R. Whan, T. Rabaa, C. Wu, D. Perera

Introduction: With the increasing mean age of the population and pressures on the health care system to increase accessibility while decreasing cost of care, there is an acute imperative to develop smarter products, devices and therapeutic approaches. At the same time, as the regulatory processes regarding combination products and devices become more streamlined, there is a great opportunity to use computational modelling for prospective design of designer dressings *cum* drug delivery devices. Here, as a first step to test feasibility of the approach we use multiphysics modelling, and a three-pronged optimization and sensitivity analysis approach, to develop and prioritize combination product/delivery device designs for *in vitro* testing. We hone a previously developed approach to directing flow direction in novel materials, using fortuitous patterns of stiffness and permeability [1,2], by accounting for encapsulation and release kinetics for active biologic or pharmaceutical cargo as well as by using click chemistry to monitor the binding state of the cargo to be delivered. In this way, we create novel, bespoke wound dressings *cum* delivery devices that signal the wearer when the bandage efficacy abates.

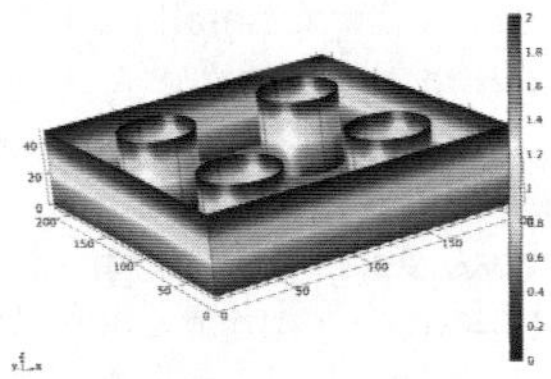

Materials and Methods: A flexible sensor array was adhered to a subject's arm over the course of a day and the range of deviatoric and dilatational stresses were monitored in a spatial and temporal fashion; these measures served as model mechanical input data for a mechanoactive combination product/delivery device. Using Comsol Multiphysics® Modeling Software and Matlab (Mathworks), one 3D module of a wound dressing and/or delivery patch was modelled. Each node of the 3D module was defined by an initial state and then exposed to a sweep of mechanical signals representing magnitudes and directions of dilatational and deviatoric stresses to which a "Band-aid" would be exposed during normal daily activities (example mesh displacement under compression depicted in inset). The state of each node was then tracked over time and favourable combinations of properties for typical daily activities were defined. In a *post hoc* step, initial properties were tuned to better represent specific healing scenarios, and a multi-module system was tested using the full array of mechanical data measured on a single subject.

Flow directing architecture Stiffness (κ), porosity (ϕ), concentration (MKT)	**Un-/Binding kinetics** intrinsic to solute and dressing material (J R-K)	**Encapsulation and release kinetics** (PZ)	**Click Chemistry** (RW) temporal tracking/visualization of binding state and activity
Concentration, *C*	Passive	Cloak properties	Clicked active
Delivery areas: $f(\kappa,\phi,\tau,\sigma)$	Active	Cargo properties	Unclicked active
Imbibing areas: $f(\kappa,\phi,\tau,\sigma)$	Hybrid		Disabled

Table 1. Independent variables that can be controlled for in 3D-printed combination wound dressings/delivery devices.

Results and Discussion: These pilot studies demonstrate the feasibility of rapid prototyping bespoke wound dressings and drug delivery devices that harness a person's natural movement to deliver healing agents in a directionally and temporally dependent manner. Having proven feasibility of the concept, ranges of variables will be defined as technical specifications for reduction to practice and follow on product development.

Conclusions: Using rapid prototyping and click chemistry, we expect the next generation of designer dressings and drug delivery devices to be customizable and 3D printable in the General Practitioner's or nurse's office. Such devices will exhibit novel functionality, from delivery of drugs and biologics including stem cells, to active collection and monitoring of wound exudate, to modulation of the wound healing cascade through spatial and temporal presentation of factors that modulate cell behaviour (migration, adhesion, proliferation, differentiation). They will signal the patient when it is time to return to the medical provider for follow up care.

Conflicts of Interest: M. Knothe Tate has a patent (U.S., EU) around the topic of flow directing materials. At the time of abstract submission the intellectual property had not yet been commercialised (pre-revenue state).

Acknowledgements: Paul Trainor and Walter Coulter Foundations.

References: [1] Knothe Tate M.L. *et al.* (2009) *Biomaterials* 30: 133-40. [2] Knothe Tate M.L. *et al.* (2011) *PLoS one* 6: e28702.

Fatigue Testing of Femoral Hip Stems – Setup Variability Examined by FEA and DOE

Danny L. Levine, Justin Hertzler
Zimmer Biomet, Warsaw, Indiana, USA

Introduction: Design verification of femoral hip implants includes an assessment of fatigue performance measured with respect to standard requirements. For experimental fatigue testing, the most widely used standards are ISO 7206-4 [1] (stem region) and ISO 7206-6 [2] (neck region), which define load criteria and detailed test specifications including design concepts for fixtures and alignment jigs. While detailed specifications help to promote test repeatability and reproducibility (R&R), the standards do not address the effect of variability in hip implant potting level and orientation, which can influence the applied bending moments and resulting peak stress. This study uses finite element analysis (FEA) and design of experiments (DOE) methods to study variations in stem testing performed using ISO 7206-4.

Materials and Methods: Simulations were performed using a straight tapered stem design. A single small implant size was selected to represent a typical worst case condition, and nominal CAD geometry was used. Geometric parameters were defined in the FE software environment to vary (from nominal) the **a)** angle in the frontal plane (α=10°), **b)** angle in the lateral plane (β=9°) and **c)** potting level measured from the head center (D=80mm) [1, 3](Figure 1). A response-surface design sensitivity study was performed using DOE tools in the FE software suite, with potting angles set at nominal ±1° and potting level set at nominal ±2mm [1]. To avoid local stress anomalies associated with modeling a potting material that is bonded to the stem, the potting material was omitted and the FE modeling approach specified in ASTM F2996-13 [4] was employed. This involves sectioning the hip stem 10mm below the potting level (at the "section level"), applying a constraint to the distal face of the extra 10mm portion, and recording predicted peak stress at the potting level.

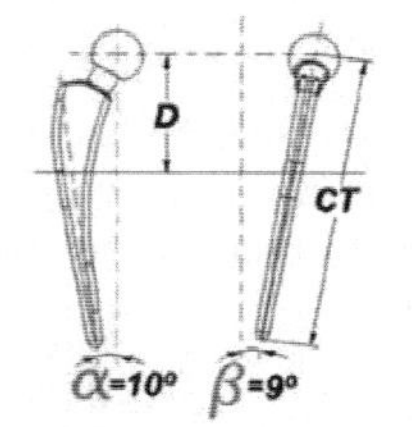

Figure 1 – Potting angles α, β and depth D [3].

Results and Discussion: With the three parameters active in the DOE, stem stress prediction at the potting level varies significantly due to the compound effect of moment arm and section property variation. Approximately 50% of the variation noted comes from the potting angles (30% from α and 20% from β) and the remaining 50% from the potting level, D. Figure 2 shows the response of predicted potting level stress to the most influential FE model input parameters, section level (potting level D +10mm per [3]) and α.

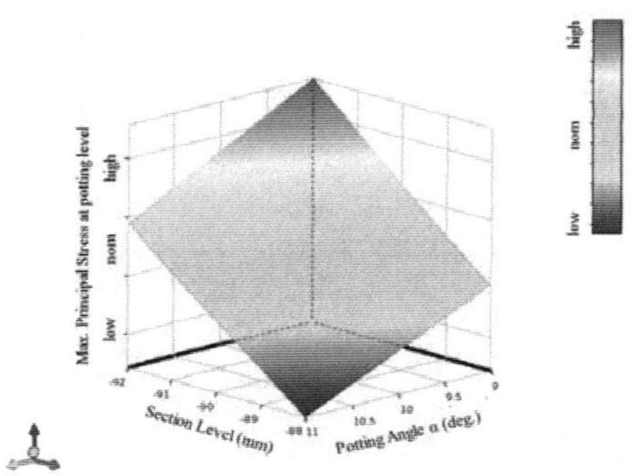

Figure 2 – Normalized potting level stress as a function of section level and angle α.

There are known simplifications in the FEA standard [4] relative to the experimental test standard [1] that the FEA is intended to support through worst case determination. Most notably, in the FE model, the stem is sectioned and rigidly constrained 10mm below the potting level, whereas the physical test specifically includes the deformable potting media. Additionally, the FEA is specified as quasi-static linear elastic stress analysis, whereas the physical test includes millions of cycles of loading in which various inelastic phenomena can occur. Though these simplifications may be appropriate for worst case determination, the extent to which the sensitivities in the model as documented here translate to sensitivities in the physical test remains to be explored.

Translational Impact: Product design decisions often rely on bench top fatigue tests performed after FEA determines a worst-case size from a product family. Conclusions from either FEA or bench top testing should reflect an appreciation of variations in setup and implications on results.

Disclosure Statement: Authors are full-time employees of Zimmer Biomet.

References:

1. ISO 7206-4, Implants for surgery — Partial and total hip joint prostheses — Part 4: Determination of endurance properties and performance of stemmed femoral components
2. ISO 7206-6, Implants for surgery — Partial and total hip joint prostheses — Part 6: Endurance properties testing and performance requirements of neck region of stemmed femoral components
3. V. Correia et al., Piezoresistive sensors for force mapping of hip-prostheses, Sensors and Actuators A, 195, (2013), 133– 138
4. ASTM F2996-13, Finite Element Analysis (FEA) of Non-Modular Metallic Orthopaedic Hip Femoral Stems

Evaluating the role of multi-channel blockage as a predictor of drug-induced cardiotoxicity using machine learning algorithms.

Jaimit Parikh, Viatcheslav Gurev and J. Jeremy Rice
IBM TJ Watson Research Center, Yorktown Heights, NY 10598

Introduction: Quantifying the risk of life threatening arrhythmias such as Torsade de Pointes (TdP), a common form of drug-induced cardiotoxicity, remains a challenge in drug development. While blockage of the hERG (I_{Kr}) channels and QTc prolongation are acknowledged markers of risk, there is a consensus that risk prediction solely based on these signals provides an incomplete picture and can lead to misidentification of drugs associated with TdPs. Examination of drug-induced blockage of multiple cardiac ion channel currents (in addition to hERG channels), that are critical in the regulation of cardiac action potential, may provide additional predictive power[2–5].

Materials and Methods: Classification of TDP risk have been examined previously using statistical methods and machine learning approaches on direct and/or indirect input metrics obtained from examination of the effects of the drugs on multiple ion channels[2,3,5]. The ratio of inhibitory concentration (IC_{50}) of the drug for multiple channel and its free effective therapeutic concentration (ETPC) (direct metrics) and features extracted from simulated action potential and calcium waveforms obtained from in-silico cardiac myocyte models for different drugs (indirect metrics) have been considered for the classification. Here, we utilize similar approaches on the 86 drugs used in the previous study[3] and constructed machine learning classifiers (Support Vector Machine (SVM) and Neural Networks (NN)) on one of three input features sets: 1] IC_{50}/ETPC, 2] blocking potencies (IC_{50}^n / IC_{50}^n + $ETPC^n$) and 3] features extracted from in-silico models parameterized based on blocking potencies. To our knowledge classifiers based on blocking potencies have not been examined previously. In addition, we also used the newly available data from the Comprehensive In Vitro Proarrythmia Assay (CiPA) working group (ion channel panel)[1], where they examined the effects of 30 clinical drugs against seven ion channels (I_{Kr}, $I_{Na,rapid}$, $I_{Na,late}$, I_{Ca}, I_{to}, I_{Ks}, I_{K1}), compared to only 3 used in previous studies.

Results and Discussion: We found that for the 86 drugs tested previously in[2] the classifier built on IC_{50}/ETPC performed the worst with an accuracy of about 62% for both SVM and NN. The accuracies were estimated using leave-one-out cross validation method. The classifier built on blocking potencies resulted in a classification accuracy of 86% based on the blocking potencies of I_{Kr} and I_{Ca} channels alone; addition of blocking potency of the I_{Na} did not improve accuracy. Interestingly, accuracy was not increased by simulating the drug-induced block of I_{Kr}, I_{Na} and I_{Ca} in an in-silico cardiac myocyte models and constructing classifiers based on the features obtained from the simulated cardiac action potential and calcium waveforms (similar to approach used in[2]). For the dataset from CiPA, we found that the classifier built on the blocking potencies of I_{Kr}, $I_{Na,late}$ and I_{Ca}, provided accuracies as high as 81% (SVM) and 85% (NN). Utilizing the blocking potencies of these three channels at three times the maximum effective therapeutic drug concentrations improved the classification accuracy to approximately 89%. If considered, the blocking potencies of other channels (I_{to}, I_{Ks}, I_{K1}, $I_{Na,rapid}$), in addition to I_{Kr}, $I_{Na,late}$ and I_{ca}, did not improve the classification and in most cases resulted in a drop in accuracy. However, if the blocking potencies of either of $I_{Na,late}$ or I_{Ca} were not considered to build the classifier, the accuracy dropped by at least 10% suggesting an important role of both the late sodium current and L-type Ca^{2+} current in TdP prediction.

Conclusion: In conclusion, our work corroborates earlier studies showing that consideration of multi-channel blockage improves predictive power over blockage of hERG channel alone. Our study in particular shows that classifiers built on indirect features (features extracted from in-silico models parameterized based on blocking potencies) provided no significant improvement in terms of classification accuracy over classifiers built on blocking potencies directly, at least for the data set tested. With the direct input of block potencies into the machine learning classifier, we find an important role for $I_{Na,late}$ in addition to I_{Ca} and hERG channel for TDP risk predictions, for the CiPA data set tested.

References:

1. Crumb WJ Jr. et al. *J. Pharmacol. Toxicol. Methods.* 81:251-62, 2016
2. Kramer, J., et al. *Sci. Rep.* 3:1–7, 2013.
3. Lancaster, M. C., and E. A. Sobie. *Clin. Pharmacol. Ther.* 100:371–379, 2016.
4. Martin, R. L., et al. *J. Cardiovasc. Pharmacol.* 43:369–379, 2004.
5. Mirams, G. R., et al. . *Cardiovasc. Res.* 91:53–61, 2011.

Integration of 3D Printed Microvilli and Sensors into a Microfluidic Gut-on-a-Chip Model

Joanne Chan, Chandni Dhamsania, Monil Ghodasra, Elana Laster, Kimberly Lo, Julie Mehta, Sarah Onimus, Mayuri Patel, Veda Ravishankar, Samuel Rombro, Yasasvhinie Santharam, Joy Wang, Jessica Yau, Wu Shang

University of Maryland, College Park

Introduction: Creating an *in vitro* model of the gastrointestinal (GI) tract that mimics the structural, absorptive, pathophysiological and microbial environment of the human gut can revolutionize drug delivery techniques and potentially replace animal testing to improve efficiency and efficacy. Current studies on the gut on a chip model consist of microfluidic channels coated with an extracellular matrix that mimic peristalsis.[1] Other studies use biofabrication to recreate the bacterial-bacterial and bacterial-epithelial signaling of the microbiome found inside the gut on a chip model.[2] However, these models fail to provide a method to quantitatively measure the effects of a drug treatment on a disease. This research incorporates 1) catechol-chitosan sensors to quantify the effects of a well-studied drug, vancomycin, on a disease, *Clostridium difficile,* and 2) 3D printed microvilli structures to mimic the absorptivities of the human gut into a polydimethylsiloxane (PDMS) gut on a chip model lined with epithelial cells.

Materials and Methods: A gut on a chip model will be constructed using soft lithography and biofabrication. The base material for the chip will be PDMS. With its high elasticity, low surface tension, and gas permeability, it allows complicated structures to be achieved using soft lithography. In the second phase, a Computer Aided Design software, Autodesk Inventor, will be used to design microvilli (Figure 1A) that will be 3D printed directly onto a PDMS chip, along with a glucose oxidase sensor. In the third phase, we will create oxygen and vancomycin sensors through layer by layer deposition of chitosan and alginate films onto a gold electrode. Caco2 cells, a line of human intestinal epithelial cells, will be cultured on the chip to test (1) the viability of the chip on its own, (2) vancomycin in the chip without *C. difficile*, and (3) vancomycin in the chip with *C. difficile*. The chip's efficiency and accuracy will be determined by comparing readings from their oxygen and vancomycin sensors to the results of *in vitro* and *in vivo* studies.

Results and Discussion: A standard curve for the glucose oxidase has been established by plotting the reaction between glucose and oxygen (Figure 1B). 3D printed microvilli structures can better mimic the absorptivity of water and nutrients found in the gut. 3D printing allows the gut on a chip model to be standardized and is a cheaper alternative for replicating the gut's microenvironment. Utilizing technology and state-of-the-art devices to replace live organisms allows researchers to improve overall human health and longevity.

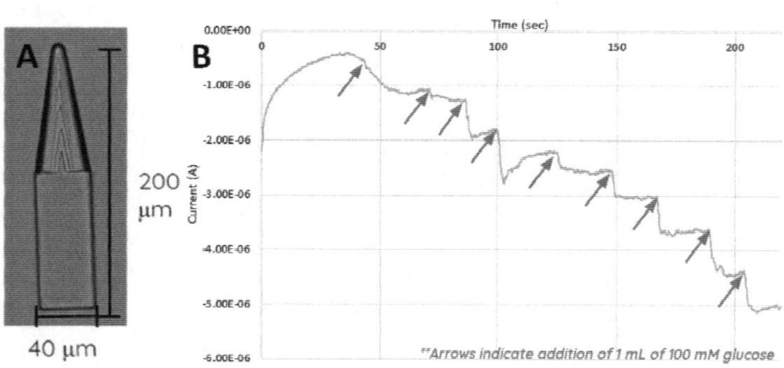

Figure 1. (A) 3-D printed microvilli structure under 100x magnification and confocal microscopy. (B) Each addition of glucose, creates a peak that represents a redox reaction between glucose and oxygen in the water to create gluconic acid and hydrogen peroxide. Glucose oxidase catalyzes this reaction and creates a current, measured in amps.

Translational Impact: By integrating 3D printed microvilli in the gut-on-a-chip model and using sensors to quantify changes in the physiological conditions of the human gut, novel drug treatments for GI diseases can be used to ensure the safety, efficacy and security of human drugs. The use of animal studies to test drugs that affect the GI tract can be significantly reduced through an accurate representation of the human gut, which can also accelerate the process for pharmaceutical and biomedical industries to approve new drugs for different GI diseases. This process can be further modified to create similar chip models for other organs, offering more efficient and reliable ways to test pharmaceutical products and improve overall human health and longevity.

Disclosure Statement: No conflicts of interest

References: 1) Luo X., Tsao C-Y., Lab Chip, 2015, 15, 1842-1851 2) Hyun J.K., Lab Chip, 2012, 12, 2165-2174

Effect of Fin Length/Shape of Stemless Humeral Component of the Reverse Shoulder Implant

Erin Ritzer, Won Joo
Engineering Department. Robert Morris University, Moon Township. PA

Introduction: The recent reverse shoulder implant designs employ shorter humeral component stems or eliminate the stem completely. They display minimal frictional fixation between the implant component and the metaphysis of the bone, which provides several advantages: preserved bone, less stress shielding, ease of revision surgery, and freedom of humoral component location due to the minimal stem length and non-complex shape. Despite its advantages, this stemless design may cause failure of the implant due to the minimal bone-to-implant contact surface, especially during early stages following surgery before osseointegration occurs. In this study, using FEA, the effects of stem length and shape on the stability of the reverse shoulder stemless humeral component was investigated.

Materials and Methods: A 3D model of a humerus and a generic reverse shoulder implant system were developed and analyzed. To investigate the effect of fin length and shape, humeral components with a baseplate radius of 18mm and various fin shape/length were developed; these included a straight fin, fins with steps, and a semi sphere with fins. Each component consisted of 6 fins of 2mm thickness and had a total length of 10.8mm, 14.4mm, 18mm and 21.6mm respectively. A press fit condition was simulated between the bone and fins, and anatomical loading was applied.

Results and Discussion: The ratios of fin lengths over the radius of the baseplate were calculated. Values for stress and micromotion were normalized by the values at 0.6 fin-to-baseplate ratio, because it showed minimum stress and micromotion. For all fin shapes, as fin length increases, the stress and micromotion decreased although the straight fin showed that the stress at ratio 0.6 and 0.8 are similar to each other (Figure 1). The results from all shapes and lengths showed that micromotion is more sensitive than the stress at the bone to changes in stem length.

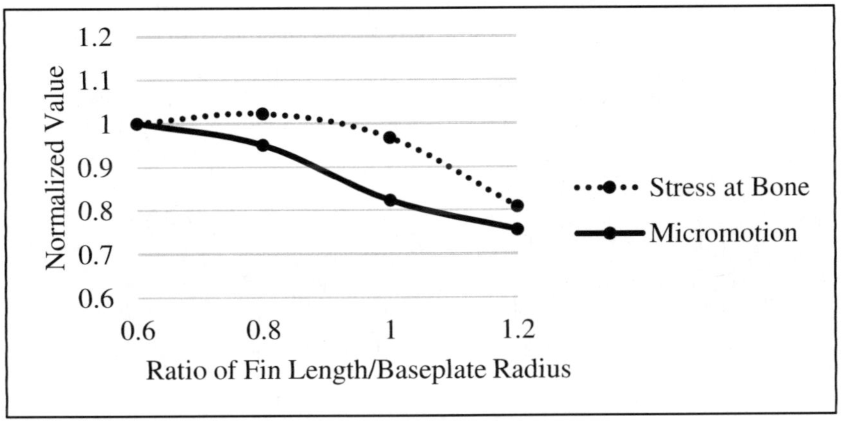

Figure 1. The graph shows the stress at the bone and micromotion of the straight fin components for different length/radius ratios. A ratio of 1 represents an 18mm fin length and an 18mm baseplate radius. Fin lengths were modified to result in the different ratios.

Translational Impact: From the data, it can be concluded that, for the case of stemless reverse shoulder humeral component, the length of the fin affects the stress at bone and micromotion. It is well known that a longer stem provides better clinical outcomes because it minimizes micromotion, and results from this study are in line with that conclusion. The results suggested that an even smaller ratio (fin length longer than 21.6mm) would be the best for stress and micromotion, but a long fin would not fit well in an anatomical humerus, and a long fin shape will eventually fall into a "short stem" category. Therefore, the results suggest that a 20-30% longer fin length than baseplate radius would be ideal for the stemless humeral component design. The various fin shapes did not have significant effects on the results and demonstrated the same trends as displayed in Figure 1.

Disclosure Statement: There are no conflicts of interest within this study.

Variation in RF Heating Characteristics of Insulated vs. Bare Metal Stents

P. Serano[1], K. Fujimoto[1], M. I. Iacono[1], L. M. Angelone[1], S. Rajan[1]
[1]U.S. Food and Drug Administration, Center for Devices and Radiological Health, Silver Spring, MD 20993

Introduction: The magnitude of radiofrequency (RF) induced heating of elongated medical implants such as stents is typically characterized as a function of the length and diameter of the device with a peak temperature increase occurring for a particular set of resonant dimensions. [1] This work aims to characterize the RF induced heating of two stent variations - conformally coated and externally covered stents - and to compare and contrast the results against traditional bare metal stents. These thin coatings have not been previously explored due to computational limitations, recently overcome by modern computing tools.

Materials and Methods: Numerical finite element method (FEM) simulations were performed to calculate the induced 1g-averaged SAR (SAR_{1g}). Devices were modeled in a tissue-mimicking gel ($\sigma = 0.47$ S/m, $\varepsilon_r = 80$) and excited by a homogeneous 128 MHz electric field incident along the length of each device. [2] The stent models for each case used a simplified geometry consisting of a hollow cylindrical tube. These simplified models have been shown previously to accurately represent stents with more complex patterns [3]. Conformally coated stents were modeled with a conformal 0.1mm coating of PTFE insulation around all surfaces. Externally covered stents were modeled with a 0.1mm layer of PTFE insulation around only the external surface (Figure 1). The peak value of SAR_{1g} located within a region 1cm surrounding the implant was used as a metric to compare across the full range of design variations. The magnitude of the resultant SAR_{1g} was normalized to the peak value across all variations. Simulations varying the implanted device's parameters were performed for 30 values of length (i.e., 5 mm to 150 mm, with step = 5 mm) and 20 values of diameter (i.e., 1 mm to 20 mm, with step = 1 mm) for a total of 600 variations for each of the three cases.

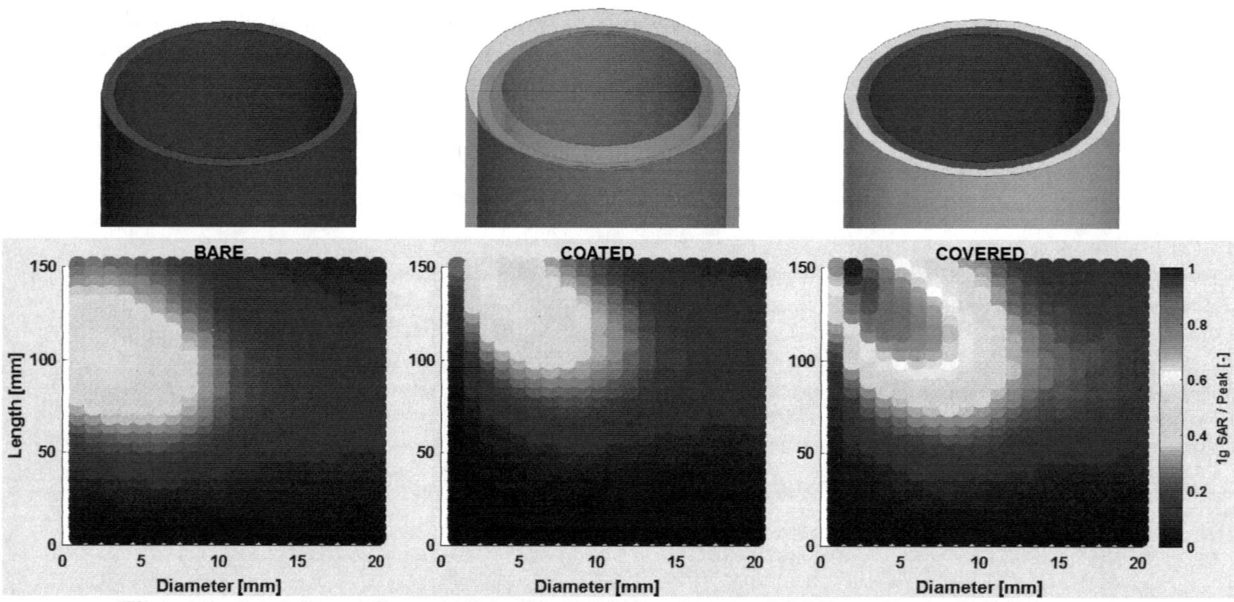

Figure 1: (top) Stent Geometries; (bottom) 1g-averaged SAR vs. Length & Diameter Bare Stent (left); Conformally Coated Stent (center); Externally Covered Stent (right)

Results and Discussion: The RF heating characteristics for the bare metal stents were as expected with the resonant length approximately equal to the half wavelength of the incident wave in the dielectric gel. The addition of the conformal insulation increases the resonant length of the device and the addition of the external insulation increases both the resonant length and overall magnitude of SAR_{1g}. These results indicate that once dielectric insulation is added to stent structures, the RF heating characteristics are no longer simply characterized by the length and diameter and further analysis of insulation thickness and relative dielectric constant must be explored.

Translational Impact: The RF heating behavior of stents varies drastically with the addition of dielectric insulation, quantified with precise computational modeling.

References: [1] Winter L., MRM (2015) [2] Bonmassar G., IEEE NEC (2013) [3] Santoro D., PLoS One (2012)

Title of Abstract: Tissue-Engineered Vascular Grafts For Use In Vascular And Reconstructive Surgeries

Gurtej Singh PhD., Kai-li Liang, John Cordero, Brody Wiles, Gabriel Klein MD., Miriam Rafailovich PhD., Nicos Labropoulos PhD, Duc Bui MD., Sami Khan MD., Marcia Simon PhD., Richard Clark, MD. Alexander Dagum MD.

Introduction: Patient receiving reconstructive surgeries may require the use of an autologous vascular graft in order to provide an extra length to the artery or vein during anastomosis to the native vessels. Additionally, an autologous vessel may also be necessary in turbocharging a graft. Currently, the standard for these procedures is to use the patient's own saphenous vein to serve as a graft. Unfortunately, this option is not always available and could result in complications related to its harvesting. Here, we employed a novel modular approach (provisional patent filed) to construct a 3-layered (mimicking tunica intima, media and adventitia layers) tissue engineered vascular graft that could potentially alleviate problems associated with existing grafts.

Materials and Methods: Crosslinked gelatin was used as a scaffold material to provide structural support to the vascular graft. For the design of the vascular conduit, a cylindrical Polydimethylsiloxane (PDMS) mold of 6mm in diameter was used. The developed lumen was then coated with fibrin and collagen type I, followed by seeding with 3 x 10^6 cells /mL human umbilical vein endothelial cells (HUVECs). Following incubation, the graft is connected to a peristaltic perfusion pump and media is being run at flow rates ranging from 3 to 31 mL/min. The tissue architecture as well as its mechanical properties were characterized through immunostaining, imaging using confocal microscopy and through rheology studies.

Results and Discussion: Endothelial cells exhibited high viability (>90%) as measured via confocal microscopy using Hoechst 33342 and propidium iodide staining. Following 7 days of incubation, the necessary structural integrity of the endothelial monolayer lining the lumen was achieved. This was proven by the formation of a well-defined network of actin-filaments, stained with Alexa Fluor 488® phalloidin stain. The diffusional permeability of the lumen was also characterized through the use of fluorescent labeled molecules such as FITC labeled dextran of different molecular weights and Alexa Fluor® 594 conjugated with bovine serum albumin. The graft was successfully operated under pulsatile flow and so leaks were observed during the time period of 36 hours. The elastic modulus and the shear stress of all the biomaterials used (collagen, fibrin and gelatin) were evaluated and the highest values were found for the gelatin (elastic modulus – 12,000 Pa, shear stress- 1000 Pa). The next steps will focus on characterization of parameters including the burst strength, fatigue strength, maturation processes, vascular remodeling both in vitro as well as animal models.

Translational Impact: We have successfully developed a novel method to generate a vascularized conduit while maintaining precise spatial control over the deposition of different biomaterials. The developed vascular graft has the potential to aid in flap based tissue reconstructions when combined with a tissue engineered construct.

Disclosure Statement: The authors have no conflicts of interest.

Engineered Outer Blood Retina Barrier for Disease Modeling of Age-related Macular Degeneration

[1]Min Jae Song, [1]Russell Quinn, [1]Roba Dejene, [2]Marc Ferrer, [1]Kapil Bharti
[1]National Eye Institute, [2]National Center for Advancing Translational Sciences

Introduction: Outer blood retina barrier (BRB) is formed by tight junctions of the retina pigment epithelium (RPE), Bruch's membrane, and choroid located in the outer region of the back of the eye. It regulates the supply of macromolecules from and to the choriocapillaris to the neural retina. Among the tissue layers in outer BRB, RPE plays a key role maintaining the homeostasis & the microenvironment around photoreceptors by regulating their viability and function. Malfunction and atrophy of the RPE cells are central causes to ocular diseases such as "wet" and "dry" age related macular degeneration (AMD). A key RPE function includes secretion of cytokines in a polarized manner: vascular endothelial growth factor (VEGF) is predominantly secreted from the basal side and pigment epithelium derived factor (PEDF) is predominantly secreted from the apical side. It is believed that RPE regulates viability and growth of choriocapillaris by differentiating cytokine secretion. Disturbance of this function can cause excessive blood vessel growth or degeneration of choricapillaris in the choroid. In wet AMD, excessive blood vessel growth, called choroidal neovascularization (CNV), breaks through RPE barrier and leaks fluid in the subretinal space. In dry AMD, atrophy of RPE cells degenerates choriocapillaris and photoreceptors. Both diseases cause severe vision loss. Currently, there is no good *in vitro* or animal model for AMD. Here, we successfully developed an engineered outer BRB using 3D bioprinting technology and human induced pluripotent stem cells.

Materials and methods: The process of the bioprinted choroidal tissue includes: 1) design structure (type of tissue, shape, thickness); 2) optimize "bioink" type; 3) choose cell concentration and cellular composition of bioink; 4) optimize printing substrate; 5) optimize media composition, volume, and media change frequency; 6) total culture time; 7) printing, and visual, histological, biochemical analysis. The optimized bioink includes iPSC-derived endothelial cells, pericytes, and choroidal fibroblasts. We printed the bioink on one side of a biodegradable polymer scaffold, and iPSC-derived RPE cell monolayer was seeded on the other side of the scaffold. We cultured printed tissues for 6~9 weeks for maturation of outer BRB. After the period of tissue culture, we performed immunohistochemistry and visualized the tissue with a confocal microscope for structural validation. We also measured transepithelial resistance for functional validation.

Conclusion and discussion: Choriocapillaris is fenestrated capillary. We tested the tissues using a fenestration marker, Plasmalemma vesicle-associated protein (PLVAP) which is also known as fenestrated endothelial -linked structure protein (FELS). We observed strong PLVAP expression in the printed "choroid" (Fig.1B). We successfully generated iPSC-derived RPE monolayer on the other side of scaffold showing that RPE monolayer expressed polarization, which indicates localization of Ezrin in apical region of RPE, and E-cadherin positive tight junctions (Fig.1A,C).

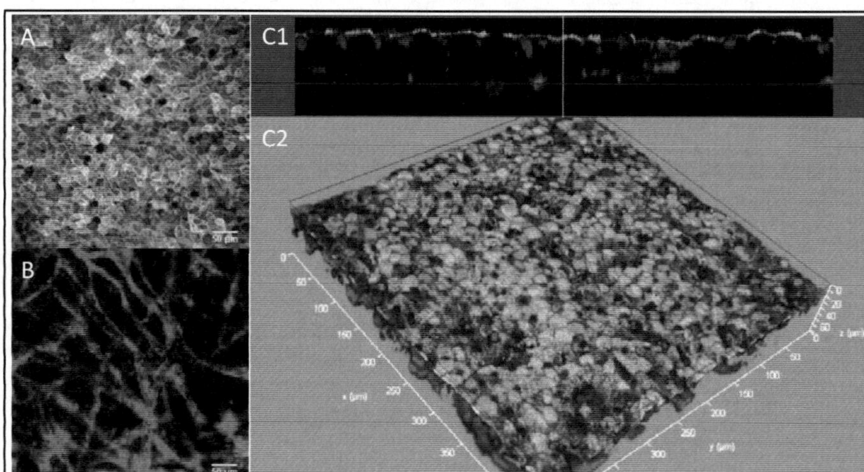

Figure 1. Engineered outer BRB on a scaffold. The tissues were fixed at 6 weeks after printing. iPSC derived RPE cells were stained with (A,C) Ezrin (green), E-cadherin (red), and Hoescht (blue). (B,C) Microvasculature, "choriocapillaris", was stained with PLVAP (red) and Hoescht (green), a marker of fenestration. Images were taken by Leica DMi8.

Translational Impact: Tissue engineering of outer BRB with patient specific iPSC-derived cells provides a novel *in vitro* system to discover underlying mechanisms of disease and provide a great opportunity to use this "disease-in-a-dish" model in high throughput drug screening platforms for drug discovery.

Disclosure Statement: There is no conflict of interest.

Modeling and Verifying Capacitance Response of Hydrogel Degradation over Enzyme Sensors

George Banis, Luke Beardslee, Reza Ghodssi
University of Maryland, College Park, MD

Introduction: This work focuses on the capacitive readout of biomaterial degradation for biosensing applications, employing finite element modeling to validate our experimental results and enhance device design. In previous studies, biomaterial films have been used as substrates to functionalize sensors for measuring specific enzymatic activity.[1] Enzyme levels outside the normal range in the small intestine can indicate a compromised exocrine pancreas, which secretes digestive enzymes from the pancreatic duct into the duodenum to aid nutrient absorption. Pathologies resulting in abnormal enzyme levels can be serious, including pancreatitis, cystic fibrosis, and adenocarcinoma. Therefore, we envision implementing sensors into ingestible capsules toward in vivo monitoring of enzymes from pancreatic secretions in the small intestine. Trypsin is one such enzyme; it is responsible for breaking down peptides at lysine and arginine residues. Here, we employ gelatin films – abundant in the aforementioned amino acids – deposited over Au interdigitated electrode (IDE) sensors to measure how the change in capacitance correlates with a decrease in film thickness from tryptic digestion. We compare this response to COMSOL simulations, in which we observe similar trends in capacitance change and can optimize designs for measuring electrical properties of these films.

Materials and Methods: IDE sensors are comprised of parylene-coated IDE sensors fabricated with Cr/Au using standard photolithography and lift-off processes. 5% gelatin in deionized water was drop-cast over the sensors and cross-linked with 0.5% glutaraldehyde for thermal stability. The capacitance of the sensors was measured using LCR meter (Agilent E4980A) controlled with LabVIEW. Measurements were carried out sequentially in air, in phosphate buffer (PB; pH 7.5), and finally in 1 mg/mL bovine pancreatic trypsin in PB at 37°C (physiological temperature) under constant stirring. Sensors tested were either coated in gelatin or uncoated. AC signals were swept from 100 kHz to 1 MHz at 50 mV in every 90 seconds. We observed films before and after each experiment and compared extracted parallel capacitance experimental results to COMSOL Multiphysics simulation results. For the latter, one pair of electrode fingers of equivalent geometry to the sensor electrodes were modeled, and coated with dielectric materials with conductivity and permittivity approximations characteristic of gelatin and parylene from previous studies.[2] Direct DC capacitance responses were calculated for films with a 100 μm average thickness or 0 μm (no film) surrounded by either buffer or air. Simulation capacitance was obtained for only two fingers in the sensor, but was scaled up accordingly for IDEs.[3]

Results and Discussion: Figure 1a is an SEM of the film over the sensor before enzymatic digestion. Figure 1b presents the simulation model, and figure 1c presents the capacitance responses of the experimental results compared to the simulation model at various conditions. Though exact capacitance is slightly offset between experiment and simulation results, the trend occurring from different conditions is preserved. This offset is most likely attributed to (a) differences in dielectric constant of gelatin when cross-linked with glutaraldehyde, as the material properties in the simulation were for uncross-linked gelatin, or (b) inconsistencies in film heights. Ultimately, this enables monitoring the change in capacitance with the degradation of the biomaterial film.

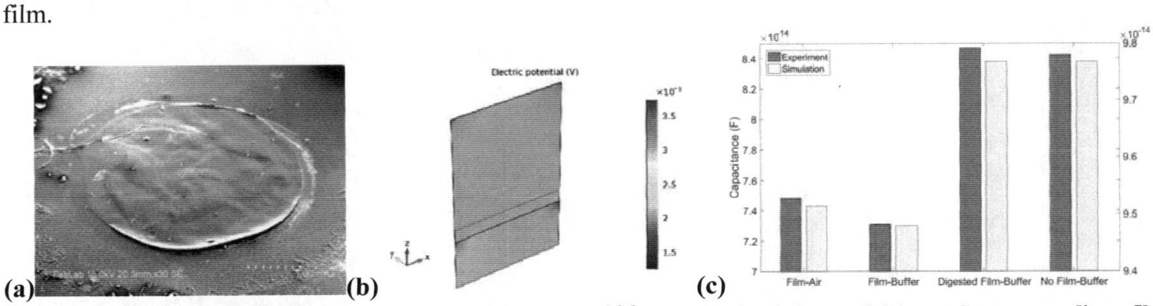

Figure 1. (a) SEM of gelatin film with an average thickness of 100 μm, (b) simulation model (top to bottom: medium, film, sensor, wafer), (c) mean capacitance from experimental degradation of film over IDE sensor and simulation results.

Translational Impact: By characterizing and verifying these responses in COMSOL, we can begin to optimize IDE design to tailor sensitivity to different initial film thicknesses for implementation in an ingestible capsule. This allows for improved process development for higher sensor throughput and approximating changes in response for varying concentrations of enzyme while simultaneously applying this model to other enzyme-specific degradable films as well.

References:

[1] Sabot, A. & Krause, S. (2002) *Anal. Chem.* **74,** 3304–3311.
[2] Fricke, H. & Jacobson, L. E. (1939) *J. Phys. Chem.* **43,** 781–796.
[3] Bilican, I., Guler, M. T., Gulener, N., Yuksel, M. & Agan, S. (2016) *Microsyst. Technol.* **22,** 659–668.

Design of a Smart-Sensing Dental Drill

Rebecca Butler, Ethan Murphy, Ryan Halter
Thayer School of Engineering, Dartmouth College, Hanover, NH, USA

Introduction: Iatrogenic injury caused by accidental perforation of the cortical bone surrounding the inferior alveolar nerve or maxillary sinus occurs in 10-40% of dental implants. No real-time sensing method is currently available to guide oral surgeons and general dentists in safe and accurate placement of a dental implant as they drill through cancellous bone towards these cortical interfaces. Significant electrical impedance differences have been demonstrated between cancellous and cortical bone and we propose a device that utilizes this impedance contrast to sense the position of a surgical drill bit during dental implant procedures. A finite element model was used to estimate the expected impedance of a drill bit moving toward an interface; these simulated findings were compared to a bench-top laboratory model of our device.

Materials and Methods: A finite element mesh was constructed in Gmsh, consisting of a 0.5mm diameter drill bit with a 1mm long conductive tip and insulated shaft, a large return electrode parallel to and 10cm away, and a bilayer sample with conductivities of 0.2 S/m and 0.0667 S/m to imitate cancellous and cortical bone, respectively (Fig. 1). Forward simulations were performed using custom electromagnetic models implemented in MATLAB to estimate absolute impedances as the needle moved in 0.5mm increments through the sample. Models with both a perpendicular and off-angle approach vector were used to evaluate how the impedance signal is expected to change with respect to the drill bit's approach to the impedance interface. Simulations were compared with data recorded from a 2 mm dental drill bit coated in Diamond-like Carbon everywhere except the distal 2mm (Fig. 2) as it moved away from the bottom of a plastic tank filled with 0.1 S/m and 0.01 S/m saline solutions. The axial sensitivity, defined as the distance away from a high impedance boundary at which the impedance changes less than 0.5% with continued translation, was calculated for both impedance boundaries in the simulation as well the drill bit experiment.

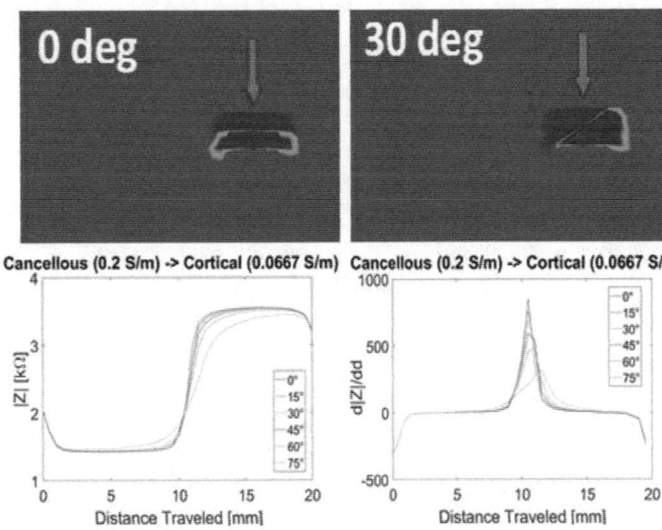

Figure 1. FEM simulation of angle of incidence (0-75°). Cancellous (dark blue) was modeled as 300% more than cortical (red) in line with previous experiments.

Results and Discussion: In the FEM simulations, impedance followed a decaying exponential in the initial 5 mm as the probe was translated from a high impedance boundary (Fig. 1). Axial sensitivity of the probe was determined to be 2 mm from the perpendicular saline (0.1 S/m)/cancellous (0.2 S/m) interface. At the cancellous (0.2 S/m)/cortical (0.0667 S/m) interface, axial sensitivity was 3 mm for a 0-30° approach angle, 3.5 mm for 45°, and 4 mm for 60°. Laboratory results matched well with simulated trends (Fig. 3), demonstrating 2.5 mm axial sensitivity at 1-100 kHz. These values indicate our prototype can be used to sense impedances at least 2.5 mm in front of the drill bit.

Translational Impact: The axial sensitivity region indicates how far in advance our device could detect an interface before a breach occurs. Further work is needed to experimentally confirm that as the drill bit passes through perpendicular and off-angle approaches to an interface the recorded impedances match the simulated trends.

Disclosure Statement: This work was completed under a sub-award made to Dartmouth College from RyTek Medical, Inc.

Acknowledgements: This work was supported by the U.S. NIH under Grant 1R41DE024938-01

References: Renton, T. Prevention of iatrogenic inferior alveolar nerve injuries in relation to dental procedures. SADJ 65:350-63, 2010.

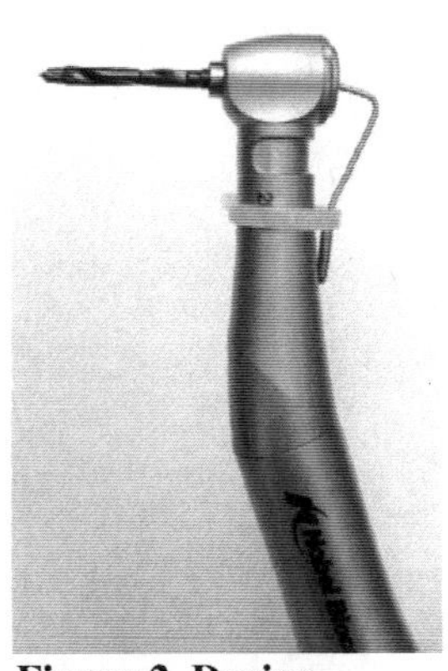

Figure 2. Device prototype, including the partially insulated drill bit.

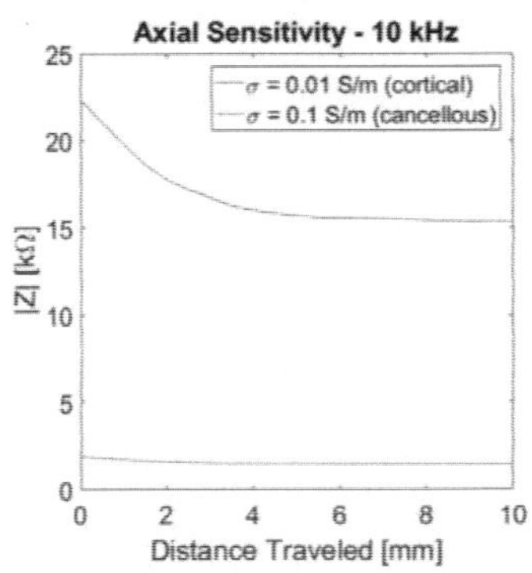

Figure 3. Absolute impedance as the drill bit was translated from the bottom of a plastic tank. A similar trend was observed in the initial 5 mm of simulations.

Biomechanical Comparison between Mono-, Bi-, and Tri-cuspid Valve Architectures

Henry Y. Chen[a,b], Sean Chambers[c], Fedor Lurie[d] and Ghassan S. Kassab[a,b]

[a]*California Medical Innovations Institute, Inc., San Diego, CA;*

[b]*Department of Biomedical Engineering, IUPUI, Indianapolis, IN*

[c]*Research Engineering, COOK® Medical, Bloomington, IN*

[d]*Jobst Vascular Institute, Toledo, OH*

Introduction: An understanding of the relationship between venous valve design and associated fluid and solid mechanical forces will undoubtedly advance prosthesis design and treatments. The objective of the current study was to compare three valve designs (mono-, bi- and tri-leaflet) and the implications of these designs on the fluid and solid mechanics of the valve leaflets. The hypothesis is that the bileaflet valve has the lowest mechanical cost, defined as the ratio of leaflet wall stress and fluid wall shear stress (WSS), for the venous environment as compared with mono- and tri-leaflet valves.

Materials and Methods: To address this hypothesis, fully coupled, two-way Fluid-structure Interaction (FSI) computational models were developed and simulated for the three types of venous valves.

Results and Discussion: It was found that the mean fluid WSS of bi-leaflet valve was generally higher than tri-leaflet valve which was further higher than mono-leaflet valve. The mean solid wall stress of bi-leaflet valve was lower than tri-leaflet valve which was further lower than mono-leaflet valve. Therefore, the mechanical cost which was defined as solid wall stress/fluid WSS of bi-leaflet valve was the lowest.

Translational Impact: The lower mechanical cost may be a reason why the bi-leaflet valve is the dominant design in the venous system. This knowledge may also provide guidance to the design of novel venous prosthetic valves and shed light on venous valve disease.

Disclosure Statement: None.

Modeling and Feedback Control of a Rapid Pressure Swing Adsorption Medical Oxygen Concentrator

Matthew D. Urich[1], Rama Rao Vemula[1], Mayuresh V. Kothare[1]

[1]Department of Chemical and Biomolecular Engineering, Lehigh University, Bethlehem, PA

Medical Oxygen Concentrators (MOCs) are a growing commercial industry due increasing demand for high purity oxygen therapies. Medical oxygen, oxygen >90% purity, is the common treatment for Chronic Obstructive Pulmonary Disease (COPD) common in elderly patients. COPD patients have a decreased oxygen uptake capacity in the lungs which makes breathing very difficult. Medical oxygen boosts the patient's blood oxygen level and mitigates the effects of COPD. MOCs are devices which produce medical oxygen from ambient air without the need for heavy, high-pressure cylinders. COPD patients typically have low physical ability, and they cannot tolerate carrying a heavy device. MOC designs focus on minimizing weight and size of devices to improve the patient experience. MOC systems are complex, nonlinear, cyclic processes which are extremely difficult to model and optimize. In this work, we present a MOC model and a feedback controller predicts the nonlinear MOC dynamics and allows the device to reject process disturbances.

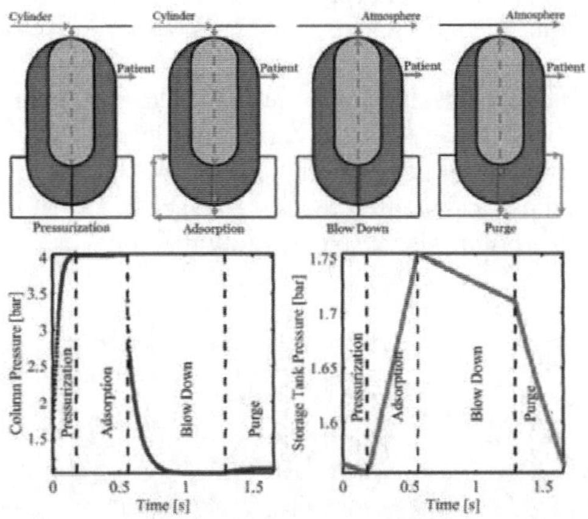

Figure 1: The MOC is a cyclic process with four cycle steps. Each step must be modeled and solved sequentially to capture the nonlinear dynamic behavior.

We model the MOC using six partial differential equations (PDEs) and 4 ordinary differential equations (ODEs) which capture the nonlinear dynamic response of the MOC unit. Because this process is cyclic, different cycle steps must be solved sequentially. We present here a framework for modelling different cycle steps and transferring solution information between cycle steps. The feedback controller is a Multi-Model Predictive Controller (M-MPC) which uses a collection of linear models to accurately represent the nonlinear model behavior. This collection of linear models is found using sub-space identification techniques. The M-MPC can manipulate the cycle steps to improve the MOC performance, and response to process disturbances which could affect the device performance. The MOC model is solved in Comsol Multiphysics, and Matlab is used for M-MPC calculations.

Modeling a MOC process is crucial for optimizing commercial designs. The MOC process is very complex and highly nonlinear, so the model is difficult to solve. We evaluate our model performance based on simulated process disturbances, and we can judge how well the M-MPC can maintain proper device performance. In our simulations, the M-MPC MOC system greatly improves the device performance.

Title: Additive manufactured FFF 3D printed osseointegrated prosthesis for a transhumeral amputation using Nylon 680

Authors: Steven Lathers[1], David Nyer, Karmella Haynes, Jeffrey La Belle[1]
Authors addresses:
[1] School of Biological and Health Systems Engineering, Arizona State University, 550 East Orange St. Tempe, AZ 85287-9709

Introduction: There are about two million people with limb loss in the United States and approximate hospital costs of $8.3 billion for amputations each year. Prosthetics can cost up to $50,000 per limb with the possibility of it not being covered under insurance. Current prosthetics are commonly attached to the user with a socket and sleeve system which cause user discomfort due to changes in stump shape throughout the day, socket shifting during use, and sweat pooling. To overcome these issues, osseointegrated prosthetics are used for direct skeletal attachment but they are made of titanium that causes stress shielding, require two invasive surgeries, and allow for a small amount of bone ingrowth. Our work presented here is a 3D printed osseointegrated prosthetic that reduces the number of surgeries, has large room for bone ingrowth, has a lower elastic modulus then bone to help prevent stress shielding, and is stronger than humeri bones.

Materials and Methods: The new prosthetic was evaluated on an Instron and manufactured on a MakerGear M2 FFF printer. Percent infill and layer height was characterized for the material using dog bone samples for infill and shear testing samples for layer height strength. The 3D printed prosthetic was evaluated in a simulated steel environment with a three and four fin design in tension and three-point bending tests, and a creep/torque out test to determine mechanical seating performance of the prosthetic. Additionally, the four fin design was evaluated in a cadaver and steel environment for three-point bending and torsion, along with initial cytotoxicity testing with human bone cells to evaluate the polymer as an implantable material.

Results and Discussion: The results below show a prosthetic that can withstand forces greater than synthetic and cadaver adult humeri bones, and fall impact forces. Additionally, we show positive cytotoxicity results and comparable fixation results between cadaver and simulated environments. The cadaver samples use a smaller size screw due to cadaver sample sizes and are compared to a simulated environment with the same scaled down screw size. Additionally, the scaled down screw is compared back to our original screw size to show the changes in strength when scaling down.

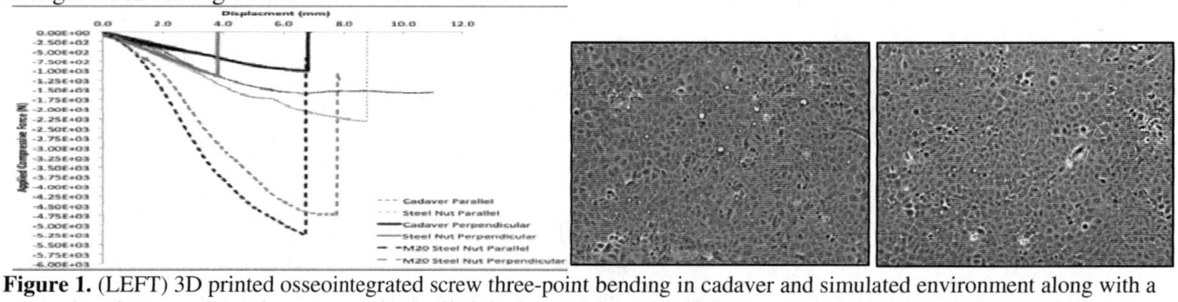

Figure 1. (LEFT) 3D printed osseointegrated screw three-point bending in cadaver and simulated environment along with a comparison between the cadaver screw size and original screw size. (CENTER) Control cytotoxicity results. (RIGHT) Cytotoxicity results for polymer compatibility. Torque, tension, previous three-point bending tests curves, and comparison table of the differences between the cadaver and simulated environment are not shown due to space requirements.

Conclusions: We present here an alternative to current titanium osseointegrated prosthetics with a thermoplastic FFF 3D printed polymer osseointegrated screw that can withstand forces greater than an average humeri bone. The screw is a one step surgical procedure for implantation with large volumes of bone ingrowth, which will allow for an increase in fixation strength, and a decrease in rehabilitation time and hospital costs. Additionally, the FFF 3D printed screw shows positive cytotoxicity results for an initial 48hour live-dead assay biocompatibility test and comparable results for the fixation strength between a cadaver and simulated environment.

Conflicts of Interest: There are no conflicts of interest with the work presented.

Impact of Patient Orientation on Pacemaker Lead Tip Heating During MRI Exams

Jessica A Martinez M[1,2] and Daniel B. Ennis[1,2]

[1] *Department of Bioengineering, University of California, Los Angeles, CA, United States,* [2]*Department of Radiological Sciences, University of California, Los Angeles, CA, United States*

Introduction: Up to seven out of ten patients with a pacemaker or an implantable cardioverter defibrillator (ICD) will likely require an MRI exam during their lifetime [1]. However, the radiofrequency (B1) induced electric field generates currents along the device lead resulting in lead tip heating (LTH). Consequently, one principal safety concern is the temperature increase that occurs at the device-tissue interface (lead tip), which may result in tissue damage. In routine clinical practice, patients with pacemakers are positioned in the scanner depending on the anatomical region under examination. A head-first (HF) supine orientation is generally preferred for scans between the head and chest, whereas a feet-first (FF) supine orientation is used for the pelvis and below. Abdominal imaging can be acquired head first or feet first. This general guideline does not consider that the induced electric field (E-field) magnitude is spatially asymmetric [2]. For example, the E-field is greatest at the left-anterior and right-posterior aspect of an ASTM phantom placed HF for circular polarized B1 fields. The E-field asymmetry suggests that patient orientation can play a role in reducing LTH. The ***objective*** of this work was to compare the LTH in five pacemakers for HF and FF orientations using a realistic implant configuration in the ASTM phantom during RF induced heating at 1.5T.

Materials and Methods: Experiments were performed for five (N=5) different pacemakers (M=3 were bi-ventricular) with an identical realistic lead path. The lead path was obtained from a whole body 3D CT scan of a cadaver with an implanted pacemaker and biventricular leads. A 3D printed mold was constructed to hold each of five different pacemakers in the identical configuration with minimal material to mitigate interference with the induced E-field. The 3D mold was placed in an ASTM torso phantom filled with polyacrylic acid with an electrical conductivity of σ=0.49 S/m and exposed to 4 W/kg SAR using a turbo spin echo sequence (TE/TR = 6.2/110ms, Flip Angle=180°) for 5 minutes at 1.5T (Siemens Avanto Fit) for both HF and FF orientations. The temperature increase (ΔT) at one or both lead tips was recorded using a fiber optic temperature probe (Lumasense Technologies) for nine different landmarks (LM) placed at isocenter. LM1 corresponded to the center of the phantom's head and LM2-LM9 were placed at 9cm intervals increasingly in the inferior direction.

Results Discussion: Fig. 1 shows ΔT for HF (Fig. 1A) and FF (Fig. 1B) for all the landmarks. Fig. 1C shows the median ΔT for the five pacemaker leads at each LM. Median ΔT values are recorded in Fig. 1D.

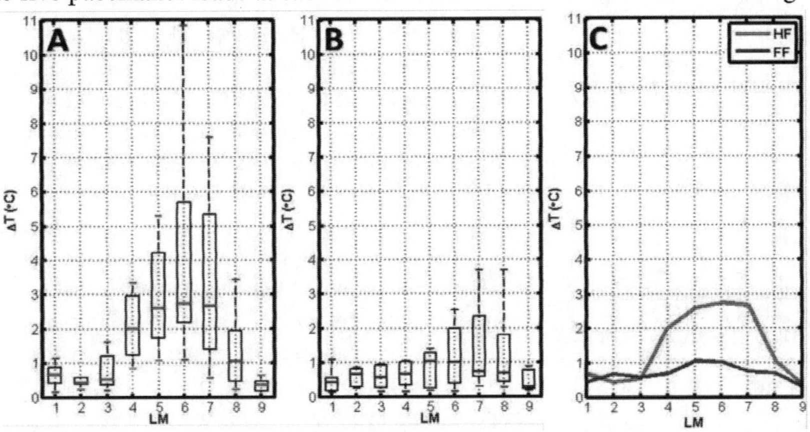

D

	HF	FF
LM1	0.7±0.3	0.4±0.3
LM2	0.4±0.1	0.7±0.5
LM3	0.5±1	0.6±0.9
LM4	2±2.3	0.7±2
LM5	2.6±3.1	1.1±2.8
LM6	2.8±3.4	1±2.5
LM7	2.8±2.7	0.7±1.9
LM8	1.1±1.1	0.7±1.2
LM9	0.4±0.2	0.3±0.8

Fig. 2 Maximum ΔT as a function of Landmark (LM) for HF and FF RF induced heating at the pacemaker lead tip for five devices. **(A)** shows the distribution of ΔT for HF, **(B)** distribution of ΔT for FF, **(C)** the median for both HF and FF and **(D)** ΔT median values.

Heating at LM1, LM2 and LM9 were similar, low (ΔT did not exceed 2C), and independent of patient orientation. However, greater values of ΔT were observed at LM3-LM8 for the HF orientation compared to FF. The maximum ΔT value was observed at LM6. ΔT was reduced by a ~2x for the FF orientation at LM4-LM7.

Translational Impact: Changing patient orientation can reduce device heating in the MRI scanner. For a conventional MRI scanner, some exams will be restricted to a HF or a FF orientation (i.e. head exams can only be procured HF). However, body regions at or near LM6 to LM8 (approximately abdominal) can be examined in either orientation. The FF orientation always exhibited lower LTH than the HF orientation for these LM. This results suggests that for exams near LM6 to LM8, patient orientation should be FF supine when the pulse generator is a left-sided implant and the scanner's pole is oriented along the head direction.

Disclosure Statement: DBE receives research support from the NIH, AHA, St. Jude Medical, and Siemens Medical Solutions.

Acknowledgements: Funding support from the Department of Radiological Sciences and NIH R21 HL127433.

References: [1] Kalin R *et al.*, PACE. 2005; 28, 326–328. [2] Nordbeck P et al., MRM 2008; 64(2):312-9.

Meeting the Needs and Challenges of a Growing Biomedical Device Industry: An Example of How Teams of Mechanical Engineers and Software Designers Can Collaborate to Optimize the Development of Various Biomedical Devices and Equipment Utilizing their CAE Simulation Expertise.

John M. Stribley[1], Joseph Formicola[1]

Caelynx[1], Ann Arbor, MI

Introduction: Caelynx is an engineering consultant firm based in Ann Arbor, Michigan with over 100 years-worth of combined Computer Aided Engineering (CAE) expertise within the company. Although Caelynx has developed a strong business portfolio working with the automotive industry, over the past 10 years there has been a gradual and increasing shift toward simulation analysis of new and existing biomedical devices and equipment. Over the past 10 years, leading biomedical design and manufacturing companies have steadily contracted with Caelynx and its partner, Dassault Systemes – an engineering software company, to conduct CAE level analysis of their products with the goal of optimizing design and enhancing performance prior to manufacturing.

Methods: The applications of various engineering software programs such as Catia, Simulia, Xflow, and CST were employed, where applicable, to solve a variety of design problems for our clients. A retrospective study of our project work-load revealed a steady increase in the percentage of hours dedicated to CAE/FEA type analysis of biomedical projects brought to our attention by new and existing clients as illustrated in the graph below.

Results & Discussion:

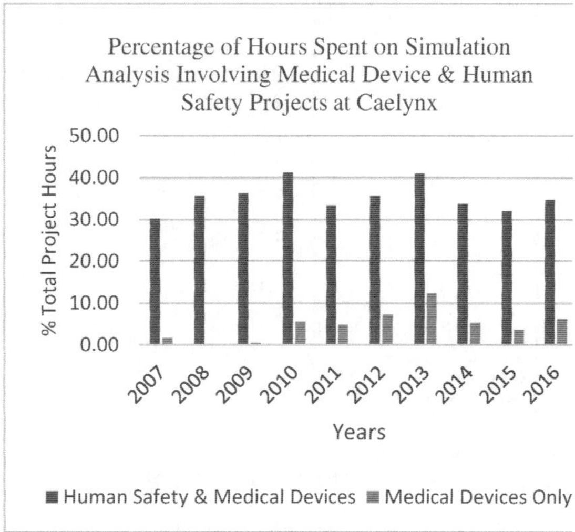

For nearly a decade, Caelynx and its partner company Dassault Systemes have seen a steady increase in CAE/FEA (Fine Element Analysis) simulation technology being applied to biomedical devices and equipment. The utilization of CAE simulation analysis in product design and testing has proved to be invaluable in the automotive, aerospace, and defense industries. Caelynx and Dassault have demonstrated that this model can be applied to the development of biomedical devices and equipment as well. Reducing costs and optimizing performance are essential ingredients in being competitive with the development and introduction of new biomedical devices and equipment to the market. The newer software platforms have allowed designers and engineers to easily go from concept to simulation to additive manufacturing in short order.

Conclusion: Growth in CAE simulations within the biomedical device and equipment industry looks robust as viewed through the lens of our company and its partners. We believe the future of the biomedical devices industry will increasingly grow to depend upon simulation analysis and additive manufacturing practices as a means of staying competitive and propelling the field forward.

Large Bioresorbable Stents for Congenital Heart Disease

Welch TR[1], Wright, J.[1].

1. University of Texas Southwestern Medical Center, Dallas, TX.

Introduction: Poly-L-Lactide (PLLA) has been investigated as an implantable biodegradable material that has more recently been investigated for biodegradable stents. An ideal bioresorbable stent would successfully treat congenital heart disease such as aortic stenosis and then disappear. Prior work was performed on a Ø3mm PLLA stent design to examine localized yielding regions of the stent. The goal of this research is to design a Ø10mmx 18mm stent 3D model and analyze the stent with Finite Element Analysis and experimental techniques for biodegradable stents used in congenital heart disease.

Materials and Methods: PLLA fibers were hot melt extruded and drawn to a 6:1 ratio for 0.25±0.01 diameter. Tensile testing of these fibers were performed on an Instron 5565. These fibers were then wound into a stent configuration. PLLA stents were analyzed by expanding at incremental balloon pressures from 0 to 8 atm. The stent size of Ø10mm x 18mm was then modeled in Solidworks 2016 and imported into Abaqus 6.13 (Dassault Systemes Simulia Corp. Providence, RI). A mesh of 252,204 elements was used for the stent model and 38,400 for the balloon model. A frictionless rigid cylinder balloon was used to analyze the expansion of the stent to a 10mm balloon diameter and a second step to simulate the stent relaxation after removal of the balloon. A nonlinear model is used imposing von Mises isotropic strain hardening yielding criterion using the material properties and stress-strain curve from the PLLA fibers. Stent hoop strength testing was performed on an Instron RX 550/650.

Results and Discussion: The mechanical properties of the 0.25 thick PLLA fibers showed Young's Modulus of 5380±26 MPa with the ultimate tensile strength 476±16 MPa and fail strain 44±26%. These material properties were used with large deformation to simulate the stent expansion and springback analysis. The stent showed localized regions of material yielding causing the stent to permanently deform into a cylindrical position. Stent validation was performed by analyzing the elastic recoil and shortening of the stent through expansion. The experimental results showed a 1.5±0.5% elastic recoil and no shortening of the stent. Stent radial testing showed a hoop strength of 0.29±0.06 N/mm.

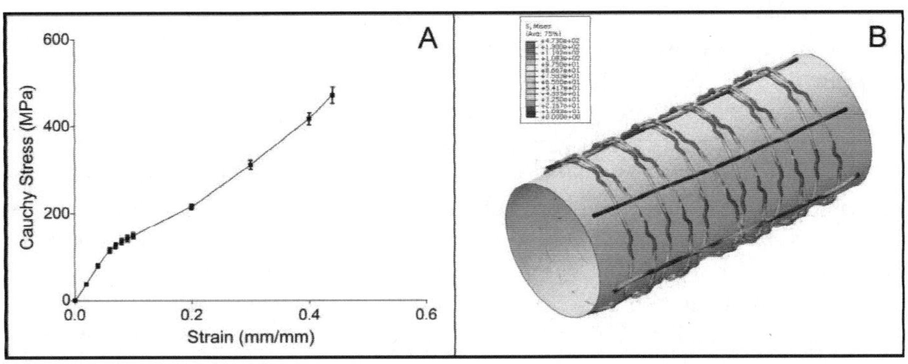

Figure 1. Showing the Cauchy Stress vs. Strain Curve for PLLA 250μm fiber (A). Showing localized yielding regions of the finite element model of the expanded stent to a 10mm balloon diameter (B).

Conclusion: A Ø10mm x 18mm bioresorbable stent has been successfully designed for congenital heart disease. Initial studies were performed on a 3 coil platform and analyzed for iliac vessels using finite element analysis. This design was modified to a 6 coil platform to successfully create a bioresorbable stent with localized plastic yielding similar to predictions in a 3 coil platform. Hoop strength testing shows the PLLA stent having a high radial hoop strength. Further studies were investigated with this approach fabricating this PLLA stent design to a 12mm diameter and compared against a metal CP Stent™ showing significantly higher radial hoop strength.

Disclosure Statement: This stent design is covered by patent US 9,155,640 B2.

Acknowledgements: Funding Source NHLBI-HL129095.

Logistics of Building Virtual Specimens for *In Silico* Biomechanics

Ahmet Erdemir[1,2], Tara Bonner[1,3], Snehal Chokhandre[1,2], Robb Colbrunn[1,3], Benjamin Landis[1,2], Erica Morrill[1,2], Tammy Owings[1], Tyler Schimmoeller[1,2]

[1]Department of Biomedical Engineering, Lerner Research Institute, Cleveland Clinic

[2]Computational Biomodeling (CoBi) Core, Lerner Research Institute, Cleveland Clinic

[3]BioRobotics and Mechanical Testing Core, Lerner Research Institute, Cleveland Clinic

Introduction: In medicine, computational modeling and simulation enables scientific discovery, facilitates engineering innovation, and supports clinical care. When virtual specimens are available, simulation-based explorations provide the means to understand healthy function of the human body and consequences of diseases, and to predict outcome of interventions. Virtual specimens can lead to virtual subjects, patients, and populations to enable *in silico* trials and for the delivery of precision medicine. In biomechanics, virtual specimens require authentic representation of their anatomical features and physiological properties to perform simulations under meaningful loading conditions. When reliable, they can replace actual specimens to reduce human subjects testing, cadaver experimentation, and animal studies. However, development of virtual specimens is prohibitively challenging when the added goal is to provide data and methodologies in an open, accessible way. This study summarizes strategies that have been adopted in our recent projects [1,2] to overcome the logistical burden of development, dissemination, and support of virtual specimens.

Materials and Methods: Strategies for building models of the knee [2] and layered-tissue structures of musculoskeletal extremities [1] are documented in their respective web sites as part of an open science driven approach. These are synthesized and provided in this document in summary form.

Results and Discussion: The activities to build virtual specimens result in heterogeneous information that need to be curated, managed, and disseminated. For any task throughout the lifecycle of virtual specimens, specifications for the task are designed and documented first. Development of virtual specimens involves data collection, data analysis, data management, model development, model management, simulations, and simulation management. *Data collection* is intended to support authenticity of the virtual specimens. It spans imaging for reconstruction of specimen-specific anatomy, tissue testing for specimen-specific material properties, and joint or organ level loading for specimen-specific validation studies. *Data analysis* steps bring raw data in a form usable for modeling and simulation, e.g., generation of meshes, identification of constitutive parameters, etc. For various development stages, automation can increase efficiency and reduce user/developer error. For example, data analysis scripts can extract model parameters directly from raw data. *Model development* steps can also be streamlined, i.e., surface representations of tissue components can be meshed and assembled, connections and interactions can be defined based on proximity analysis, and model files can be exported in various simulation software formats. Quality assurance of data is established by reproducibility tests at the tissue or organ level. Similar strategies for modeling quantify uncertainties in virtual representation of individual tissue components, e.g., their geometry. Version control is implemented at all stages. *Data, model,* and *simulation management* platforms and workflows associate heterogeneous data types and support traceability of models and derivative data to their origin. Document generation (user or developer), document management, and dissemination of data and models provide support for user engagement. An open development approach and free and open source dissemination ensure utmost outreach to the biomedical community. The use of free and open source software tools for data analysis and model generation, software agnostic approach for virtual representation, and cloud computing support enable anyone to build upon the capabilities of virtual specimens.

Translational Impact: Sample strategies to tackle logistical challenges of building virtual specimens can assist the biomechanics community in effective utilization of existing models and development of organ representations.

Acknowledgments: This study has been supported by USAMRMC (W81XWH-15-1-0232) [1] and NIGMS, NIH (R01GM104139) [2]. The views, opinions and/or findings contained in this document are those of the authors and do not necessarily reflect the views of the funding agencies. The authors are thankful for contributions of all past and current collaborators, see project sites for more details.

References: [1] Reference Models for Multi-Layer Tissue Structures @ https://simtk.org/projects/multis. [2] Open Knee(s): Virtual Biomechanical Representations of the Knee Joint @ https://simtk.org/projects/openknee.

Real-time MRI Motion Tracking of the Cardiac Cycle in Breath-Held, Normal and Heavy Breathing

F. Scott Gayzik[1,2], Craig A. Hamilton[1], Ashley R. Weaver[1,2]
1. Wake Forest School of Medicine, Department of Biomedical Engineering
2. Virginia Tech-Wake Forest Center for Injury Biomechanics

Introduction: Modern imaging techniques offer the ability to quantitatively capture processes within the body *in-vivo*. Digitally tracking movement of the heart and lungs in relation to the surrounding skeletal structures as a result of respiration can provide key data for the design and risk assessment of implantable cardiac devices such as defibrillators and pacemakers, which require lead placement within the heart's chambers. Real-time MRI offers a unique modality for quantifying *in-vivo* the dimensions, orientation, and location of the heart structures, great vessels, lungs, diaphragm, abdominal contents and the rib cage at different inspiratory volumes and rates.

Materials and Methods: A pilot study was conducted with an MRI-compliant healthy normal male volunteer (Age, height, weight: 26, 183 cm, 75 kg) under with three MRI scans collected at different inspiratory rates: 1) Breath-held scans were collected under 20 sec of suspended respiration, 2) normal breathing was instructed as typical tidal breathing, and 3) heavy breathing was instructed as maximum effort with an emphasis on using the diaphragm to "belly breathe". Scanning was conducted in the sagittal plane, using a Siemens Skyra 3T scanner with an 18-channel body array coil and the table's spine coil. A single-shot TrueFISP pulse sequence was employed. The FOV was 400x237 mm, with a pixel size of 3.125 mm^2 and thickness of 10 mm. The frame rate was 6.5 frames/sec, thus a temporal resolution of 150 msec. Following scan acquisition, the apex of the heart was tracked relative to the sterno-xiphoid junction using ImageJ software [2].

Results and Discussion: Figure 1 shows still images during the cycle. The right panel shows the motion of the tracked location. Average breathing frequencies were 0.34 and 0.44 Hz for tidal and heavy breathing respectively. Higher frequency data in Figure 1 represents the cardiac cycle. Resultant in-plane peak relative motion of the heart, for the heavy breathing was 62.7 mm (majority of motion in the cranial-caudal direction), vs. 22.3 mm for tidal breathing (factor of 2.75x). No motion outside of the heart beating was detected in the breath held scan. Heavy breathing exhibited a ramping effect and less control on rhythm (COV of period, *T*, was 21% vs. 4.6% for tidal breathing).

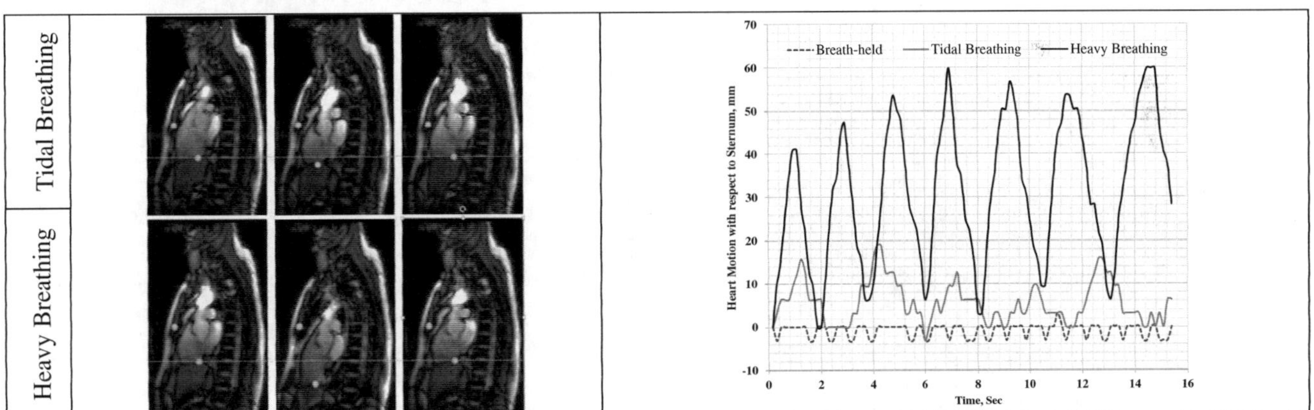

0 • *T*	1/2 • *T*	*T*

Figure 1. Left, reat-time MRI of heart cycle noting tracked locations and initial position. Above, time history of tracked point.

Translational Impact: This pilot study indicates the potential for real-time MRI to provide quantitative in-vivo data for organ tracking. Imaging was conducted in a single mid-sagittal plane only but more planes are feasible. The technique could prove valuable as a design input for implantable devices and is well suited for validation of internal motions within computational human body models. The protocol could be applied to other body regions.

Disclosure Statement: The authors have no disclosures to make.

Acknowledgement: The work was supported by Wake Forest University School of Medicine

References: [1] Quick, H.H. et al. J. Magn Res Img. 2002 Jun: 15(6): 710-5. [2] Rasband, W.S., ImageJ, U. S. NIH, Bethesda, Maryland, USA, http://imagej.nih.gov/ij/, 1997-2015.

A Smartphone Stethoscope and Application for Automated Identification of Innocent Still's Murmur

Yogesh Karpate[1], Sukryool Kang[1], Robin Doroshow[2] and Raj Shekhar[1]
1 Sheikh Zayed Institute for Pediatric Surgical Innovation, Children's National Health System Washington DC.
2. Division of Cardiology, Children's National Health System Washington DC.

Introduction: Primary care physicians (PCPs) often lack the skills to distinguish Still's murmur from pathological heart murmurs. This leads to approximately 800,000 children being referred to pediatric cardiologists each year for evaluation of heart murmurs in the United States. The murmur is ultimately diagnosed as innocent Still's murmur in approximately 90% of these children. These unnecessary referrals and associated tests cost the healthcare system over $450 million annually, and are a source of avoidable anxiety in children and families while waiting to see a pediatric cardiologist.

Our objective has been to develop a novel technology that could identify Still's murmur with high accuracy at the point of care to reassure PCPs in their decisions regarding referral to a specialist.

Materials and Methods: We have created a novel stethoscope that connects to a smartphone and developed a custom application that records heart sounds and analyzes them for the presence of Still's murmur. The stethoscope records heart sounds with a quality comparable to that of a commercial electronic stethoscope. The analysis algorithm has been developed and tested using an in-house murmur library with clinically documented diagnoses (119 Still's murmurs, 84 other innocent murmurs, and 502 pathological murmurs). The recordings were made using an electronic stethoscope (3M Littmann Model 4100). Seventy percent of the murmur library was used to train the algorithm, while the rest comprised the test set. Our algorithm offers a true positive rate (Still's murmur identified correctly) of 89% with a false positive rate (pathological murmur misidentified as Still's) of 3%. We have also preliminarily evaluated our algorithm on the heart sounds of 27 children (4 Still's murmurs, 5 normal heart sounds, and 18 pathological murmurs) recorded by our mobile stethoscope and found the murmur identification accuracy to be 92%. These preliminary results show that the diagnostic accuracy of our algorithm is similar whether the recordings are made using our mobile stethoscope or the Littmann device.

Figure 1: The proposed stethoscope coupled with a smartphone is shown as following **(A)** and corresponding Receiver Operating Curve (ROC) for the proposed analysis algorithm **(B)**.

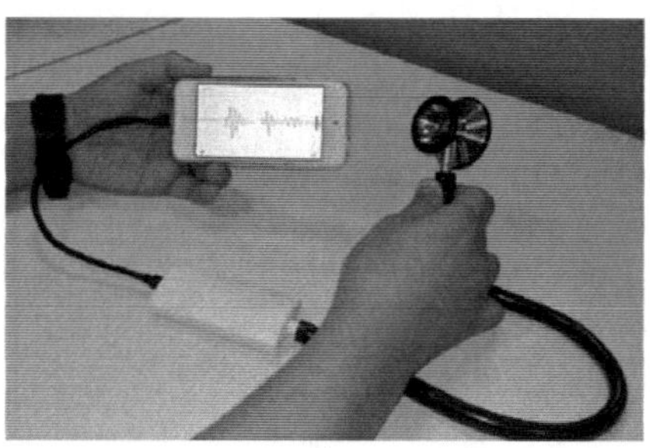

A

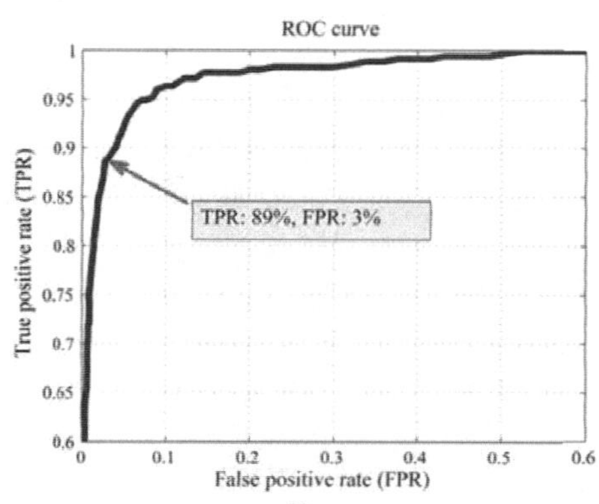

B

Results and Discussion: The described point-of-care mobile technology could automatically distinguish Still's murmur from all other murmurs with high accuracy.

Translational Impact: The technology could lower the current high rate of specialist referrals associated with Still's murmur and reduce the related financial and emotional costs. Future directions will include further improvement of the technology and validation through multi-center clinical trials

Disclosure Statement: There are no conflicts of interest associated with the study.

Tongue Computer Interface

Kevin Kerr, Nicholas Marjanovic, Ricardo Aranda, Hananeh Esmailbeigi
Department of Bioengineering, University of Illinois at Chicago, Chicago, IL

Introduction

It is currently difficult to proficiently use a computer and navigate the internet without the use of a computer mouse or trackpad. These devices are inoperable by millions of people with limited motor function of their upper extremities.This can be the result of a wide range of neurological diseases, disabilities, and injuries. For this large population, current assistive medical devices are expensive, bulky, and difficult to operate. Our team addresses these problems by creating a wireless device that allows patients with with limited hand mobility the ability to control a computer with the touch of their tongue.The tongue-computer interface (TCI) replaces the standard hand operated mouse with a complete range of motion in the X and Y-directions, via the use of a wireless, tongue controlled joystick and push button.

Materials and Methods

The Tongue Computer Interface consist of a module composed of a custom made printed circuit board (PCB) that is directly connected to a joystick and pushbutton. The PCB consist of a Atmel 32u4 Microcontroller and a MDBT40 Bluetooth LTE chip. Its measures roughly 2 cm by 1.5 cm. The microcontroller is programmed to receive a change in resistance from the movement of the joystick. When these movements are registered by the microcontroller, it transmits this movement over a bluetooth connection moving the cursor simultaneously with the joystick. The pushbutton follows the same operational path as the joystick. Currently our method is similar to that of an orthodontist, molds are created of the patient's mouth using dental alginate and plaster paris. After the plaster is hardened the use of dental thermoplastic creates a retainer consisting of the an exact mold of the upper teeth and palate.The device is then placed within the upper palate of the retainer allowing the device to enter the mouth. Our current prototype works as an alternative to the computer mouse. In order to compare these two modes of communication an experiment was designed using of a simple 6 button visual interface. The test subject was called by a random number generator to move the mouse cursor to the designated button, the random button generator repeated this ten times per test. Three test were completed per subject. This procedure was performed using a computer mouse with both the right and left hand, and the TCI using both the tongue and pointer finger.

Results and Discussion

The results from this experiment were promising, when using our device with the tongue compared to the pointer finger to control the mouse cursor, the average times per trial were similar (Figure 1). However, when comparing our device to the use of a traditional mouse we were roughly 4x slower (Figure 2). With reiterations of our design we believe we can get this difference to at least 2x slower than a traditional mouse. Currently the comfort of having our device in the patient's mouth is not ideal, as we continue to develop a more comfortable model the average time per trial will decrease.

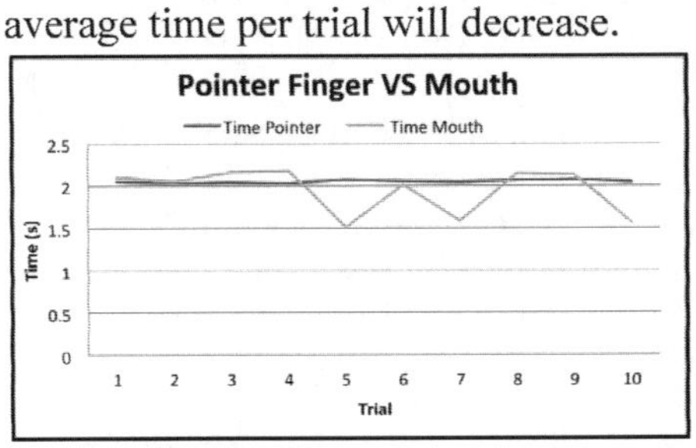

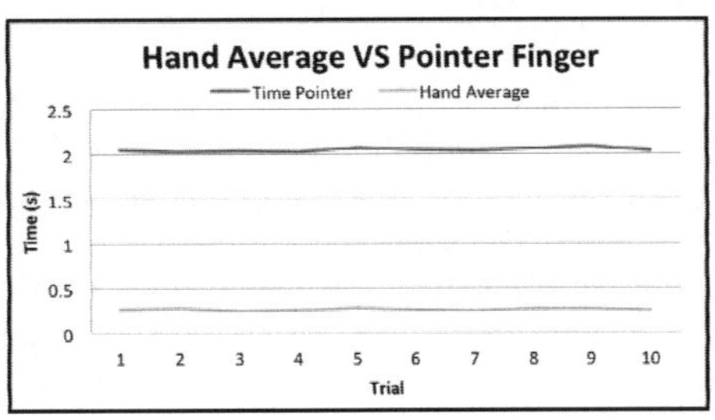

Figure 1 (Left) shows a graph comparing the average time per trial for the tongue vs the pointer finger.
Finger 2 (Right) shows a graph comparing the average time per trial for the use of the TCI vs a traditional mouse

Translational Impact

In conclusion, the device was developed by our team to assist patients with limited upper extremity mobility by giving them the ability to maneuver a computer cursor. Currently a rough prototype has been developed and the data recorded has yielded promising results. As we continue to develop our prototypes the efficiency of the device will increase as well as the possible uses for the TCI. Our aim is to develop a device that is affordable, user friendly, and comfortable for patients. We believe that our device will open new doors to patients who were once unable to operate a computer due to neurological conditions and injuries.

Non-rigid Registration of Computational Phantoms for Personalization and Population Coverage

Bryn A. Lloyd[1], Alessandro Alaia[1], and Niels Kuster[1,2]
[1]IT'IS Foundation, Zurich, Switzerland, [2]ETH Zurich, Switzerland

Introduction: Human body models are used in a variety of applications, including vehicle safety, power deposition from mobile devices such as phones, radiation from imaging techniques, such as CT/PET or dosimetry and thermal effects from MRI. In most of these cases, computational phantoms were created from full body medical image data by segmenting tissues and organs with different function and material properties, such as fat, muscle, bones or blood vessels. The Virtual Population (ViP) is a library of such models, segmented from MRI of volunteers [1]. While the ViP models already cover several extremes of the population variability, it is far from representative of all dimensions of body shape variability, such as obesity, age, gender, and organ level differences.

In this work, we present a method capable of registering a high-resolution anatomical model into arbitrary body shapes provided in the form of (partial) image segmentations or surface models. The technique deforms the template computational phantom, e.g. from the ViP library, to patient specific data. This can significantly reduce the effort required to generate new or personalized computational human models.

Materials and Methods: Our method is composed of three main stages. The first step consists of a state-of-the-art multi-resolution mesh registration algorithm for morphing the overall shape of the template model to the target. The outer surface of the template is decimated to create a hierarchy of mesh resolutions. The registration is started at the coarsest resolution and gradually adds details by moving up the hierarchy. During the second step the thickness of the subcutaneous fat (SAT) layer in the target model is reconstructed. The surface enveloping all internal tissues (we call this the internal skin) is extracted by peeling off the skin and SAT layer. The internal skin is morphed to accommodate changes in posture and size, while keeping the proportions approximately constant. The bones are registered semi-rigidly inside the new internal skin. In the third step, all soft tissues and organs are registered by interpolating the deformation field in the space between the bones and the internal skin using the harmonic extension.

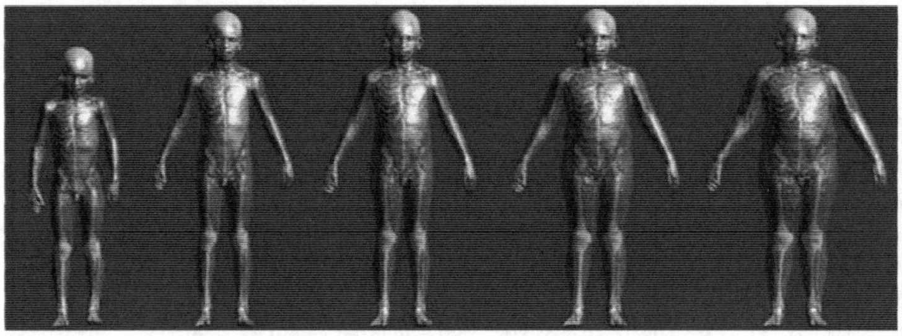

Fig. 1: Anatomical model THELONIOUS registered to several models obtained from the UMTRI child shape statistical shape model [3].

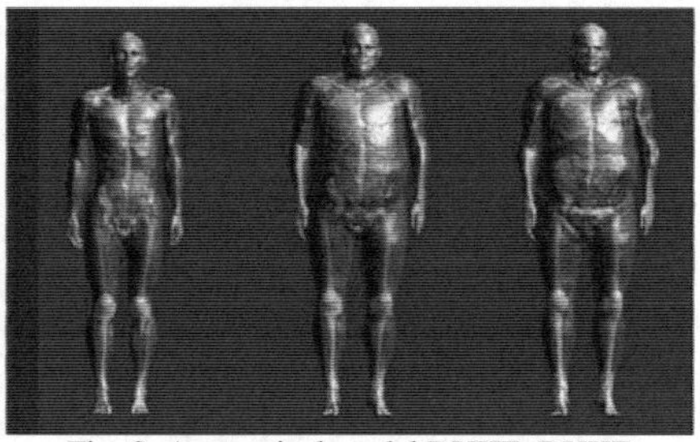

Fig. 2: Anatomical model DUKE, DUKE registered to FATS and the original FATS.

Results & Conclusions: For all experiments we have used v3.1 ViP models with more than 300 tissue structures as template models. We have experimented with publicly available surface datasets obtained from laser scans (e.g. [2], [3]). Figure 1 shows THELONIOUS registered to a child model from www.childshape.org [3] with different BMI parameters in the range 15-30 kg/m^2. As can be seen, the SAT layer is reconstructed realistically, keeping the internal skin nearly unchanged. Similar tests have been done using ELLA and DUKE for the CAESAR dataset, with very promising results. In all cases the multi-resolution surface registration step produced good matches, with minimal interaction. To verify our anatomy inspired heuristics for registering the internal tissues to a target model without using any information on internal structures, we registered the ViP models DUKE to FATS, allowing to compare estimated bone or soft tissue shape and position with ground truth (Fig. 2). While the approach is surprisingly good at predicting the internal structures, additional constraints e.g. via image or mesh registration of bones and/or specific organs is expected to provide even more accurate personalization of the template model.

Acknowledgements: This work was supported by CTI (Project: NEUROMAN CTI 25290.1 PFLS-LS). **References**: [1]. Gosselin M-C. Phys Med Biol 2014;59:5287–5303. [2]. CAESAR. www.sae.org/standardsdev/tsb/cooperative/caesar.htm. [3] www.childshape.org

Monitoring Facial Nerve Function with Digital Analysis: a Non-contact EMG Equivalent

Yan Xu, PhD[1], Divya Bhatnagar, PhD[2], Susan M. Fiore MS [3], Minh Hoai Nguyen , PhD, [1] Miriam Rafailovich PhD, [1] Raphael P. Davis MD, [3] , Duc Bui, MD[4], Sami Khan, MD [4], Richa Verma, MD [4], Alexander Dagum, MD, [4]

[1] College of Engineering and Applied Science, Stony Brook University, Stony Brook, NY, 11794

[2] New Jersey Center of Biomaterials, 145 Bevier Road, Piscataway, NJ, 08554

[3] Department of Neurological Surgery and the Neurosciences Institute, Stony Brook Medicine, Stony Brook University, Stony Brook, New York, 11794

[4] Department of Plastic Surgery, Stony Brook Medicine, Stony Brook University, Stony Brook, New York, 11794

Introduction: Digital Image Skin Correlation (DISC) is an inexpensive imaging technique that can provide physicians with both diagnostic and prognostic data for management and treatment of acoustic neuroma's (vestibular schwannomas,VS) , and other incidents affecting the facial nerves. DISC is non-contact and non- invasive and yet provides information similar to that obtained via electromyography regarding facial muscular enervation. Hence it can be used for early detection of pathologies, such as VS, as well as in cosmetic applications, such as botulism (BTX-A) injections.

Materials and Methods: The technique is simple and easily implemented anywhere, even in the patients home, using an iphone camera and sending the images to the physician from processing. Two sequential images of slight facial motion are obtained, mounted on a head support that we developed. Two studies were conducted to demonstrate the clinical application of DISC. The first involved the detection of VS and monitoring post surgical patient recovery. Two sequential digital images of slight facial motion (smile) were obtained from 29 patients with VS (with and without surgery) and 14 healthy volunteers. The images were processed with DISC to produce spatially resolved vector maps of the muscular displacement. Comparing the patients' "normal" functional side (non-tumor involvement) to the "affected" side (tumor involvement) the locus and impact on facial enervation was quantified.

A second, 12 months, two stage, double blind, study was conducted to determine the usefulness of DISC in tracking paralysis and predicting the locus of botulism injections. 10 Caucasian women ages 30 to 60, were recruited and injected in two stages with 25 U of BTX-A in the glabellar region. One group was injected using standard protocol, according to "surgeons experience", where the location of the muscle was determined by physical examination and injections were performed symmetrically. The second group used a DISC guided method: baseline images were taken prior to injection and the locus of maximum tension was determined from the vector images (figure 1 right). Tension was defined as the region where maximal changes occurred in the gradient to the vector. Guided region did not employ symmetry arguments and identified each injection locus separately. In all cases, pictures were taken in follow-up visits at days 3 and 7 and then at weeks 2,3,4, 8, 10, 12, and 24.

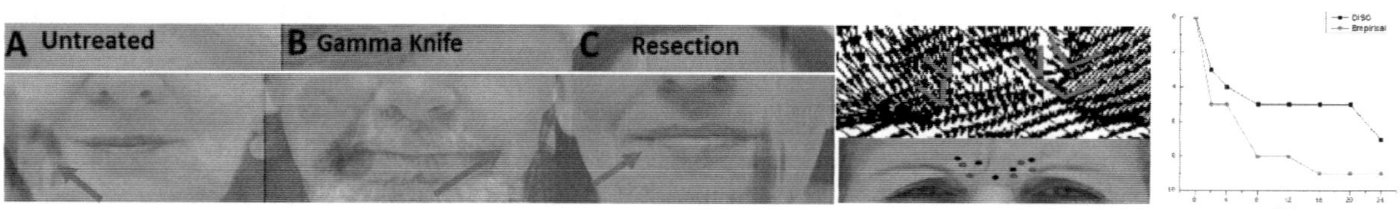

Figure 1: Left panel: Overlay of DISC generated color map on the deformed image of (A) untreated/no surgery patient (B) Gamma Knife treated patient (C) Resection treated patient. The arrows point to the location of the tumor. **Right panel**: (D) Vector map generated by DISC prior to BTX injection. Location of empirical injections (black) and DISC (red) guided are overaalyed on the glabellar region. (E) Survival curve plotting fraction of patients with more than 50% paralysis following BTX injection via DISC (black) and Empirical methods (red).

Results and Discussion: Comparing the patients' "normal" functional side (non-tumor involvement) to the "affected" side (tumor involvement) we found that .non- surgical VS patients had facial asymmetry that was five times larger than the control group, (P< 0.0001). The side with the tumor is red, exhibiting hyperkinesis, relative to the normal side. Patients who underwent traditional surgical or gamma knife resection of the VS, had asymmetries that were 10 and 15 fold higher , respectively, than the control group (P< 0.0003), where the affected side was now hypokinetic. In the case of the BTX study, the results (survival curve) clearly show that DISC is an effective method to track paralysis, and in most cases agrees with FLO-11 analysis. The results following DISC guided injections were far more effective than those from empirically determined injections. The paralysis was deeper and lasted 24 weeks vs 8 weeks.

Translational Impact: DISC has been shown to be effective in early screening and detection of VS, at a significantly lower cost than MRI, the current method. DISC is effective in tracking paralysis following BTX injection, and using DISC as a guide for the injection site, allows for a significant decrease in dosing, which minimizes off target effects.

Conflict of Interest: None

Discrete Frequency Infrared Imaging as a Tool for Stainless Histopathology of Clinical Biopsy Samples

Saumya Tiwari[1,2], Jai Raman[3], Vijaya Reddy[4], Rohit Bhargava[1,2]

[1]Beckman Institute for Advanced Science and Technology, [2]Department of Bioengineering, the University of Illinois at Urbana-Champaign, Urbana, IL [3]Oregon Health and Science University, Portland, OR, [4]Rush University Medical Center, Chicago, IL

Introduction: Histopathology is the gold standard for disease diagnosis in many diseases. However, there are several limitations of the current techniques, for example, stains are susceptible to experimental variations. In addition, while using immunostains, one must know the target well in advance. Oftentimes, these processes require multistep preparations, suffer from non-specific backgrounds, require expensive probes and as several studies have reported, have interobserver variability during interpretation of the stains[1]. Using infrared spectroscopy, a digital stain can be obtained without the need to use any chemicals or dyes on the sample[2]. While previously this technique was limited by the speed of image acquisition, new developments in the field have made it possible to image large image sections (>1cm^2) within minutes. Through this project, we explore the sources of variations in the IR images and demonstrate an application of discrete frequency IR imaging to achieve contrast and identify histology in unstained biopsy sections.

Materials and Methods: Anonymized formaldehyde fixed and paraffin embedded endomyocardial biopsy sections were used in this study. These samples were deparaffinized by washing in hexane for at least 12 hours. The samples were imaged using Agilent Hyperscan Discrete Frequency Infrared Imaging system. In order to ensure that the biochemical features(metrics) used for differentiation of histologic classes were independent of the patient to patient variations, analysis of variance (ANOVA) was performed. For ANOVA, the histologic class label was considered as fixed factor while patient id was considered as a random factor. For each metric, the p-value was calculated for the null hypothesis that the samples at all levels of the nth factor are drawn from the same population. Bayesian classifier was developed for differentiating between normal tissue (myocardium), accumulation of collagen in tissue (fibrosis) and tissue damage (cell death) using selected metrics.

Results and Discussion: ANOVA analysis showed that in most of the metrics, the difference between histologic classes was the biggest source of variation, and patient to patient differences were not significant (figure 1). Bayesian classifier was developed on the full set of metrics and a subset of metrics sufficient for classification was identified. These metrics were confirmed to be among the top 50 metrics identified in ANOVA. These features were sufficient to differentiate between histologic classes in a cardiac tissue as shown in figure 2. The band at 1038 cm^{-1} corresponding to C-O stretching indicated the distribution of carbohydrates in the sample while 1092 cm^{-1} (phosphate stretching vibrations) was used for visualization of nucleic acids and 1236 cm^{-1} band was used to identify the distribution of Amide III containing moieties. The pixel level accuracy achieved with this approach was greater than 90%.

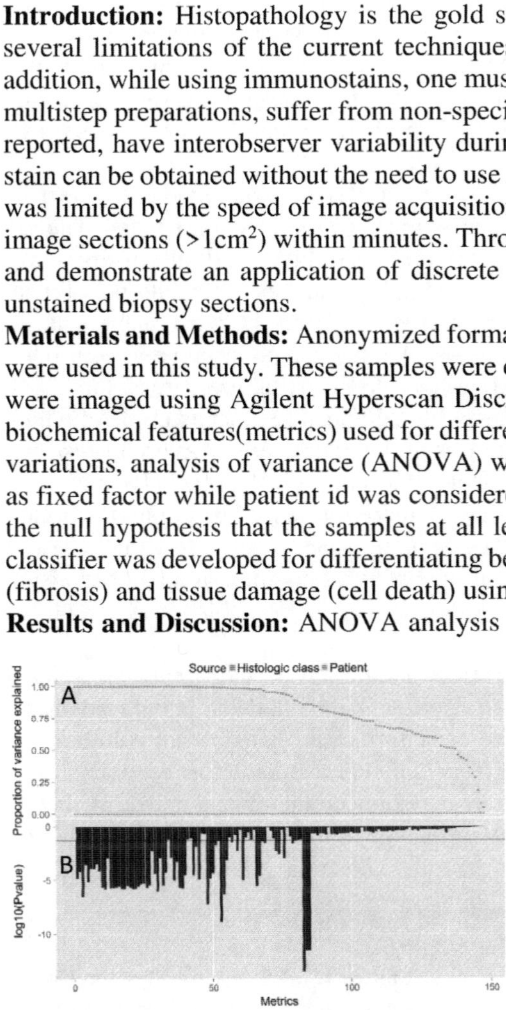

Figure 1: (A) Proportion of variance explained by source of variation (B) log10(pvalue) of source "histologic class". Horizontal line in (B) represents p-value of 0.05.

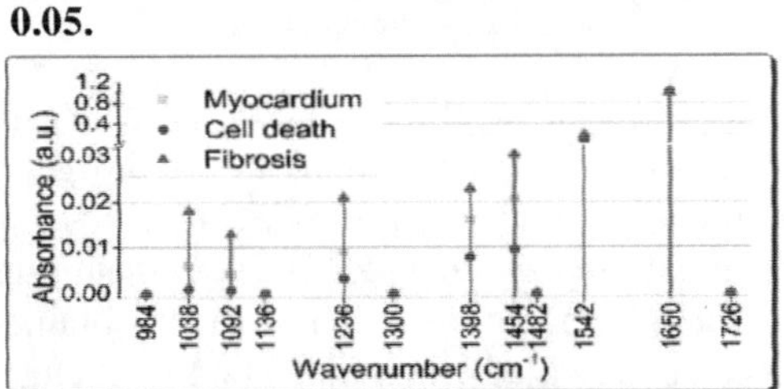

Figure 2: IR bands used for differentiation between histologic classes using Bayesian classifier

Translational Impact: Combining ANOVA and supervised classification ensures reproducibility and accuracy when new patient samples are analyzed. Through this project, we identify metrics that can enable high-speed digital histology in cardiac biopsy samples which can be readily translated to clinics.

Disclosure Statement: No conflict of interest

References: 1. Shanes, J. G. *et al.* Interobserver variability in the pathologic interpretation of endomyocardial biopsy results. *Circulation* **75,** 401–405 (1987).

2. Fernandez, D. C., Bhargava, R., Hewitt, S. M. & Levin, I. W. Infrared spectroscopic imaging for histopathologic recognition. *Nat. Biotechnol.* **23,** 469–74 (2005).

Postural Influence on Thoracoabdominal Organs of 5th, 50th and 95th Percentile Male Subjects

Ashley A. Weaver, Katelyn Greene, James P. Gaewsky, F. Scott Gayzik
Virginia Tech-Wake Forest University Center for Injury Biomechanics, Winston-Salem, NC

Introduction: Changes in posture, such as sitting down or lying supine induce alterations of the skeleton and soft tissue that impact the morphology and position of organs within the abdominal region [1]. Understanding the extent of these postural changes on thoracoabdominal organs is crucial in the design of implantable medical devices and in device evaluation with computational human body modeling [2,3]. Preceding studies have demonstrated the influence of posture on relative organ location and rib coverage [2,4,5]. This study extends the investigation of postural changes to include thoracoabdominal organ volume and morphology through development and examination of 3D models of the 5th, 50th and 95th percentile male subjects in both the supine and seated postures.

Materials and Methods: Medical images of the three males (5th, 50th, 95th percentile) were obtained in the form of supine and quasi-seated thoracic CT scans and both supine and seated (fully upright) thoracic MRI scans. The CT scan data was utilized to segment skeletal structures including the clavicles, vertebrae, and ribs using semi-automated tools. Segmented bones were then placed in the supine and seated MRI scans and manually positioned. Solid organs of interest (lungs, heart, kidneys, liver, and spleen) were segmented from the MRI scans using a combination of thresholding, dynamic region grow and manual techniques. All segmented structures were post-processed to make fine adjustments that resulted in smoothed stereolithography (STL) structures, free from holes. This process created six 3D anatomical models (three subjects in supine and seated postures), an example of which is shown in Fig. 1a. Volumes of the various organs were then calculated from the 3D models. A paired t-test was used to analyze the significance of volume changes among the subjects. Morphological organ changes were quantified by conducting an automatic best fit alignment and 3D comparison of the reference object (supine organ) to the test object (seated organ). An example of the aligned seated and supine organ is shown in Fig. 1c. The maximum and minimum deviations experienced by each organ represent the extent of compression and expansion experience by organs in the supine versus seated postures.

Results and Discussion: The volumetric thoracoabdominal results indicated that there is no statistically significant change in volume ($p>0.05$) between the supine and seated positioning among the 5th, 50th and 95th percentile males. Organ morphology changes were quantifiable and the deviation results were normalized with organ volume. The normalized deviation value was calculated using the root mean square error (RMS) of the deviation between supine and seated organs according to RMS/(Supine Organ Volume)$^{1/3}$. The liver experienced the most significant change in morphology given its volume while the kidneys experienced the smallest change in morphology (Fig. 1b and 1c).

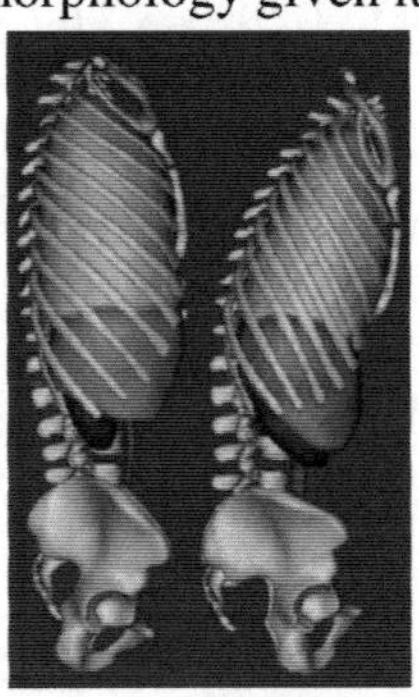

(a)

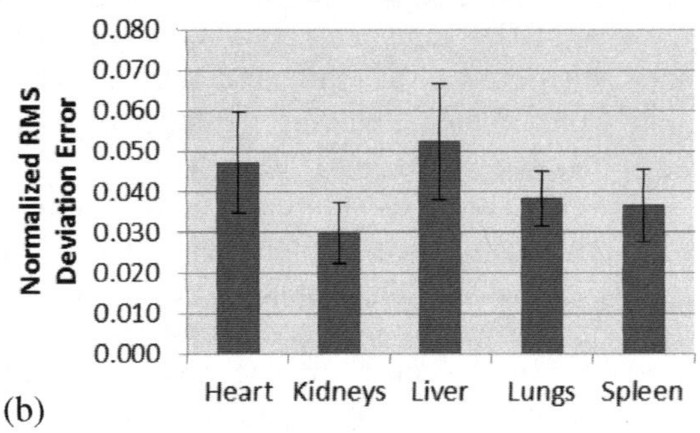

(b)

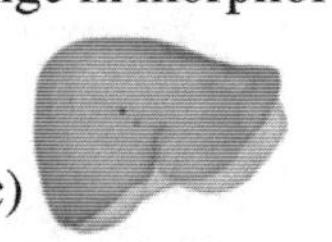

(c)

Figure 1. (a) Side view of 3D model of supine subject (left) and seated subject (right). (b) Graph exemplifying morphology change resulting from posture change. (c) Supine and seated liver overlapped to demonstrate morphology change.

Translational Impact: Quantifying the movement and morphology of the heart, lungs, and other soft tissue organs in relation to the skeletal structures in a seated versus supine posture has important implications in the design of medical devices such as implantable cardioverter defibrillators or pacemakers which require placement of leads within the heart chambers. Anatomical variation within the population must also be considered when designing devices for individuals of different sizes and shapes (i.e. 5th, 50th, versus 95th percentile). The results obtained suggest that anatomical variations along with postural changes and the corresponding organ morphology changes should be considered in future device designs and computational human body models.

Disclosure Statement: No conflicts of interest are associated with the study.

References: [1] Howes MK. Ann Adv Automot Med. (2013) 57: 209-224. [2] Lafon Y. J Biomech. (2010) 43: 1269-1278. [3] Hayes AR. Ann Adv Automot Med. (2013) 111-121. [4] Hayes AR. Comput Math Methods Med. (2013) 1-12. [5] Beillas P. Stapp Car Crash J. (2009) 53: 1-21.

The Cardiac Outcomes Risk Assessment (CORA) Counselor: Design of a Personalized LVAD Patient Decision Aid

LC Lohmueller, JF Antaki

Introduction: The decision by a patient to receive a left ventricular assist device (LVAD) requires awareness of the severity of their heart failure and potential changes in lifestyle, as well as the risks of adverse events. This information can be difficult to convey and most materials available online are from device manufacturers themselves. The goal of this work was to investigate the needs of potential users and create a first-of-its-kind personalized patient decision aid.

Methods: Interviews, patient-physician observations, and surveys were used to design a wireframe for presenting both general and personalized prognostic information for patient decision making. Data analysis includes Bayesian Search to derive insights into patient attitudes, descriptive analytics, and Tree Augmented Naïve Bayes for personalized predictive models.

Results and Discussion: Patient surveys revealed a relationship between data literacy and interest with health engagement. Different visualizations of prognostic data were tested for understanding and implicit bias. These prognostic visuals were combined with common patient questions and patient value capture to create a prototype decision aid (Figure 1).

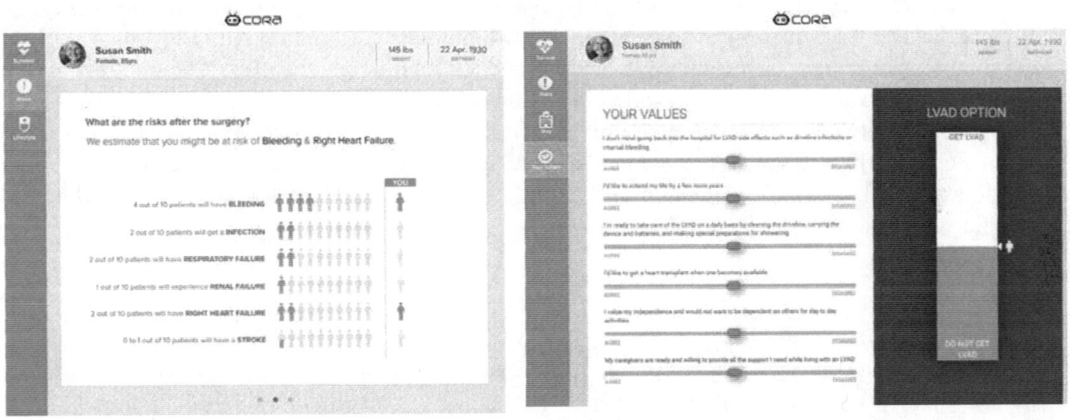

Figure 1. Wireframes of prognostic information and patient value capture in the prototype CORA patient counselor.

Translational Impact: The CORA patient counselor is first of its kind to provide decision making support through a mix of text, visual, and personalized information. It will improve shared decision making for patients and their caregivers when considering an LVAD and, we hypothesize, will improve patient satisfaction and outcomes, post-implant.

Disclosure: There are no conflicts of interest, real or perceived, to disclose.

Acknowledgements: This project is funded by the NHLBI R01 HL122639 grant. Lisa Lohmueller is a trainee of the Cardiovascular Bioengineering Training Program (T32HL076124-10) at the University of Pittsburgh.

Mechanical Evaluation of Peripheral Artery Stents

Kaspars Maleckis[1,2], Paul Deegan[1], William Poulson[1], Cole Sievers[1], Anastasia Desyatova[1], Jason MacTaggart[1], Alexey Kamenskiy[1]

[1]Department of Surgery, University of Nebraska Medical Center, Omaha, Nebraska

[2]University of Nebraska-Lincoln, Lincoln, Nebraska

Introduction: High failure rates of Peripheral Arterial Disease (PAD) stenting appear to be associated with the severe environment of the femoropopliteal artery (FPA) that bends, twists and axially compresses during limb flexion. The inability of certain stent designs to accommodate these deformations may contribute to FPA injury and result in device and treatment failure. The goal of this study was to perform bench-top evaluation and comparison of currently used PAD stents under the most severe deformations experienced by the FPA during limb flexion.

Materials and Methods: Twelve Nitinol stents (Absolute Pro, Supera, Lifestent, Innova, Zilver, Smart Control, Smart Flex, EverFlex, Viabahn, Tigris, Misago, and Complete SE) were quasi-statically tested under axial and radial compression, bending, and torsion at 37° C. Quantitative and qualitative head-to-head assessments were performed for stent force-strain behavior, stiffness, and geometrical shape under each deformation mode.

Results and Discussion: Mechanical tests demonstrated significant differences in force-strain, stiffness, and buckling behaviors between the stents. Viabahn, Tigris, Misago, and Supera were the least stiff stents under axial compression (7.0, 6.6, 31.4, and 44.7 N/m axial stiffness, respectively), while Smart Control, Smart Flex, and Zilver were the stiffest (575.3, 534.4, and 235.9 N/m, respectively). Viabahn and Tigris also had the lowest bending stiffness (3.0 and 0.1 N/m, respectively) while Smart Control had the highest stiffness (105.4 N/m). Under radial compression Supera, Complete SE, LifeStent, and Viabahn were the stiffest (295.4, 892.8, 840.25, and 760.0 N/m radial stiffness, respectively), while Smart Control and Zilver had the lowest radial stiffness (211.0 and 273.2 N/m, respectively). Viabahn, Absolute Pro, and Tigris were the least resistive to twist (2.2, 9.4, and 13.4 μN·m/° torsional stiffness, respectively), while Supera, Smart Control, and Smart Flex had the highest torsional stiffness (959.2, 85.7, and 73.4 μN·m/°, respectively). Several stents experienced global buckling and diameter pinching at physiologic deformation levels.

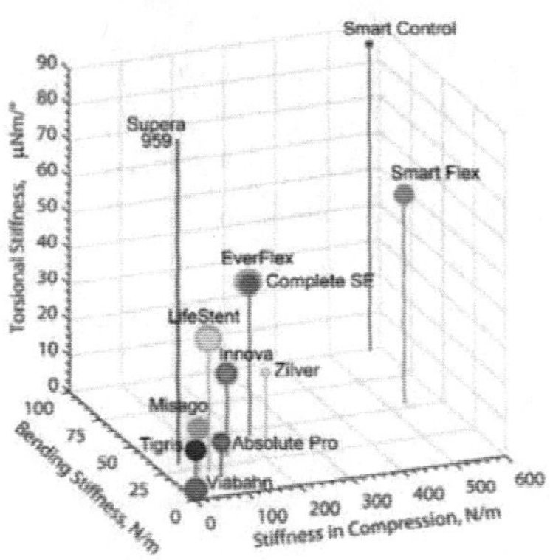

Figure 1. Comparison of bending stiffness (N/m), stiffness in axial compression (N/m) and torsional stiffness (μN·m/°) for all stents. Size of the markers represents radial stiffness scaled down by a factor of 25. Torsional stiffness of Supera was 959 μN·m/°.

Conclusions: Stent design significantly affects mechanical behavior. Though none of the 12 PAD stents demonstrated superior characteristics under all deformation modes, there are certain design features that may help stents better accommodate limb flexion-induced compression, bending and torsion. Though it is yet to be determined which of these deformation modes have higher clinical importance, results of the current analysis can inform computational models for development of new stents with improved mechanical characteristics. These results may also assist with patient and lesion-specific device selection during clinical decision making.

Disclosure Statement: Authors claim no conflicts of interest.

Thermal Imaging and Modeling of the FDM Printing Process and Its Influence on Stem Cell Differentiation

Yuval Shmueli[1], Kuan-Che Feng[1], Adriana Pinkas-Sarafova[1], Dilip Gersappe[1] Marcia Simon[2],PhD, Jake Lindberg[1], and Miriam H. Rafailovich[1], PhD
[1]Department of Materials Science and Engineering, Stony Brook University, NY, 11794
[2]Department of Oral Biology and Pathology, School of Dental Medicine, Stony Brook University, 11794

Introduction: Fused deposition modeling (FDM) is a rapidly growing method for device fabrication. The technique is inexpensive and the product, such as bone inserts, dental devices, can be printed directly from CT scans or impressions, and hence directly tailored for the individual. An important question then arises, whether devices that are manufactured from the same polymer to meet the same specifications, are actually equivalent, or further testing is required to determine if , when in contact with tissue, any differences arise at the cellular level.

Materials and Methods: Substrates for cell culture were either printed or molded from the same PLA filament. The printing process was imaged using an infra-red, FLIR camera and the thermal distribution was modeled using Mathematica. Dental pulp stem cells (DPSC) were plated on both molded and printed substrates without additional chemical induction factors. Differentiation was monitored using RT-PCR and SEM of the mineral deposits.

Results and Discussion: Computer modelling of the thermal distribution within the FDM filament was in excellent agreement with thermal images, which indicated that the fiber surfaces were in fact much colder than the nozzle temperature, where the temperture were a funciton of the nozzle geometry and the extrusion rate (Figure 1). As a result incomplete fusion as well as sharkskin roughness was introduced on the sample surface, which, as can be seen in figure 1, can have a profound effect on the differentiation pathway.

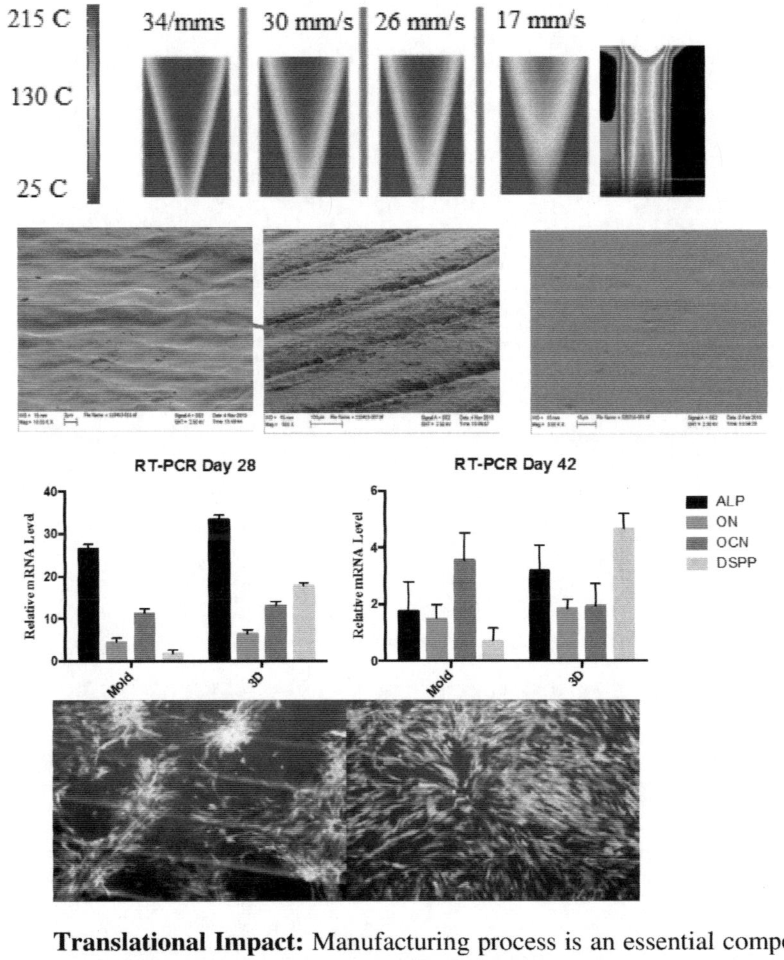

Figure 1: Top panel: Simulation of thermal distribution as a function of the extrusion speed for an FDM printer. Actual thermal image obtained during printing. **Middle panel**: (upper) SEM images of the surface from an FDM printed PLA substrate, with the section in the box magnified. Note multi scale roughness from incomplete fusion and sharkskin. SEM image of a molded PLA sample, with no remarkable features. (lower) RT-PCR analysis of dental pulp stem cells plated directly on molded and printed substrates in the absence of dexamethasone, after 28 and 42 days. Note upregulation of OCN on the molded sample and DSPP on the FDM printed sample. **Bottom panel:** Confocal image of the DPSC stem cells 11 days after plating on printed (left) and molded (right) PLA substrates. Note the cells are uniformly distributed on the molded sample, and appear to migrate into the rough features produced by incomplete fusion during the FDM printing process.

Translational Impact: Manufacturing process is an essential component in determining the function of a device. Computer modeling of the printing process allows better control of surface function.

Acknowledgement: We gratefully acknowledge support from NSF-Inspire grant, DMR-1344267.

Conflict of interest: None

The Role of Judgment in the Assessment of Model Credibility

Markus Reiterer, Medtronic, PLC, Corp. Core Technologies, Minneapolis, MN – 55432

It can be expected that over the next decade computer models and simulations (CM&S) become increasingly more important to provide evidence for regulatory submission. If CM&S provides critical information for a submission, we need to able to trust the conclusions drawn from the model. This is especially important when the simulation makes predictions that lay outside of the accessible validation domain or the comparator is subjected to a large experimental error or uncertainty. ASME V&V 40 describes a risk-informed credibility assessment framework, which is very helpful to analyze the applicability and the predictive power of a model. The document describes a 3 step process, where 1st the risk for a specific application is defined, then 2nd a rigorous verification and validation (V&V) procedure is executed, and 3rd the model credibility is assessed. The V&V step is highly technical, but 1st and 2nd step require human judgment, as there will be most of the time a lack of fundamental understanding and incomplete information available.

Office of Science and Engineering Laboratories

U.S. FOOD & DRUG ADMINISTRATION

Center for Devices and Radiological Health, 10903 New Hampshire Avenue, Silver Spring, MD, 20723

The Office of Science and Engineering Laboratories (OSEL) serves as the Center of Devices and Radiological Health (CDRH) research arm to support the assessment of the enormous breadth of devices that the Center regulates, from MRI scanners and neural stimulation devices to pacemakers, cancer treatments and other lifesaving products.

Ensuring readiness for emerging and innovative medical technologies

The pace of innovation and introduction of new technology in medical devices continues to accelerate. It is imperative for the FDA to anticipate trends to ensure that it can appropriately assess the safety and utility of new medical products when they are presented to the FDA.

A great example is the introduction of 3D printing, which is now in common use in medical devices. By the time that the FDA started to receive applications involving 3D printing, we knew the most important questions that needed to be answered, resulting in faster clearances and confidence that the devices were effective and safe for use.

Developing appropriate evaluation strategies and testing standards

Standardized testing methods have enormous benefits for both industry and the FDA, and therefore patients. The use of methods that are both well characterized and broadly accepted is a vital part of providing industry with predictable and science-based processes for the introduction of new medical products. FDA can reduce the time to review applications, accelerating the process of timely access to life-saving medical products.

Creating accessible and understandable public health information

The research that OSEL conducts results in a high number of peer reviewed publications each year. The information is publicly available and used by industry, patient advocacy groups, academia and other stakeholders to promote public health initiatives.

Delivering timely and accurate decisions for products across their life cycle

For some advanced technologies and product areas, OSEL acts as a scientific consultant to our FDA colleagues in the approval and clearance of novel devices. This is an integral function of the review process which ensures that patients in US have timely access to the broadest possible portfolio of products to treat diseases and improve the nation's health.

www.fda.gov

Case Study: Facilitating innovation in cervical intervertebral body fusion devices

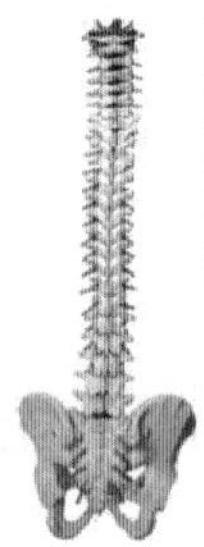

The FDA has a vast store of prior data on cervical intervertebral body fusion devices (CIBFDs), with about 2 new 510(k) applications received every week. In collaboration with Stanford, we are collating these data and making them publically available both to facilitate FDA reviews and to aid industry in product design.

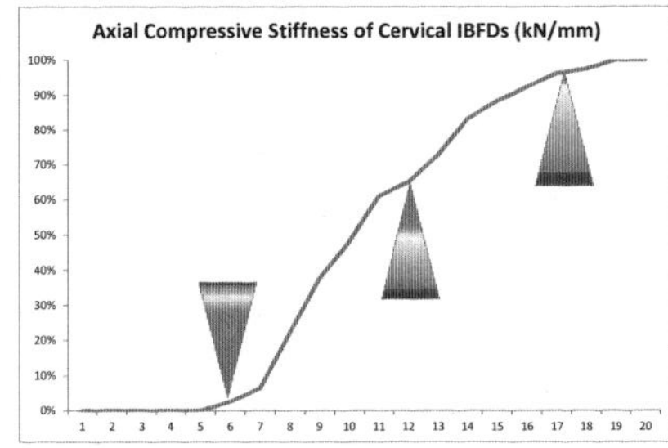

Peck, J. H., Sing, D. C., Nagaraja, S., Peck, D. G., Lotz, J. C., & Dmitriev, A. E. (2017). Mechanical performance of cervical intervertebral body fusion devices: A systematic analysis of data submitted to the Food and Drug Administration. *Journal of Biomechanics, 54*, 26-32.

Case Study: Virtual Imaging Clinical Trial for Regulatory Evaluation (VICTRE)

This research aims to demonstrate that computational modeling can play an increasingly predominant role in the regulatory assessment of imaging products. *In silico* imaging trials may allow regulators to make efficient decisions, expediting patient access to high-quality, safe, and effective devices.

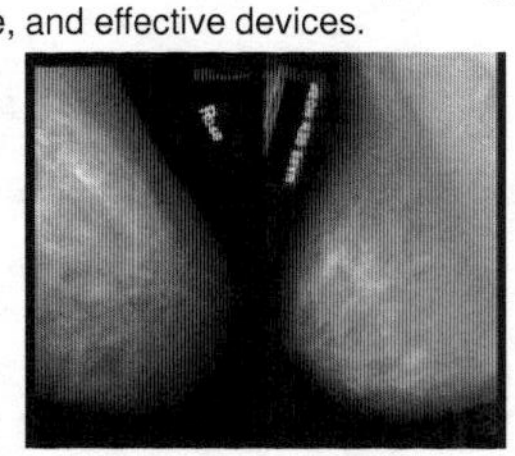

Full-field digital mammogram of a patient next to an *in silico* image of a computational model of anatomy (simulated image on the left).

Viceconti, M., Henney, A., & Morley-Fletcher, E. (2016). In silico clinical trials: how computer simulation will transform the biomedical industry. *International Journal of Clinical Trials, 3*(2), 37-46.

Case Study: Virtual Population and Multimodal Imaging Based Detailed Anatomical (MIDA) Model

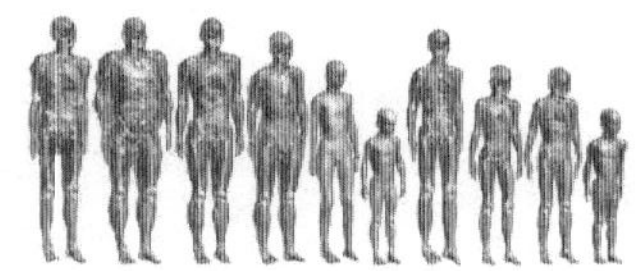

Developed by the FDA and IT'IS Foundation, these anatomically correct whole-body computer models are used for electromagnetic (EM), thermal, and computational fluid dynamic simulation. They have been used in FDA marketing applications since 2007.

Gosselin, M. C., Neufeld, E., Moser, H., Huber, E., Farcito, S., Gerber, L., ... & Cherubini, E. (2014). Development of a new generation of high-resolution anatomical models for medical device evaluation: the Virtual Population 3.0. *Physics in medicine and biology, 59*(18), 5287.

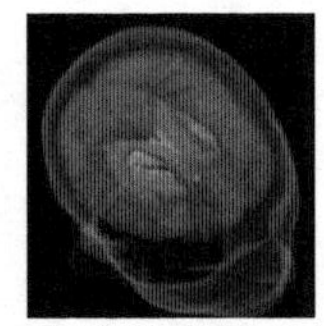

140 total model downloads; 7000 views; 26 independent citations

Iacono, M. I., Neufeld, E., Akinnagbe, E., Bower, K., Wolf, J., Oikonomidis, I. V., ... & Pruessmann, K. P. (2015). MIDA: a multimodal imaging-based detailed anatomical model of the human head and neck. *PloS one, 10*(4), e0124126.

Case Study: Providing public health information about nickel released from devices

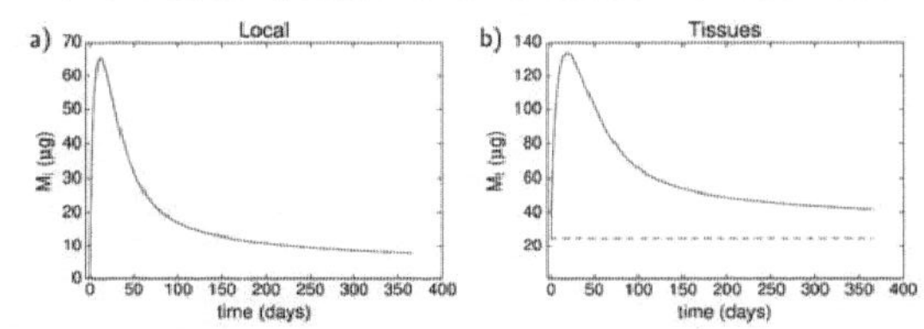

Predicted nickel content in the local a) and systemic b) tissue compartments. The local tissue nickel exhibits a maximum value of 65 µg 12 days after implantation. The maximum nickel systemically in tissue occurs a week later at day 19. In both compartments, substantive elevation in nickel levels is predicted even at one year following device implantation. The dotted line in b) represents the baseline value predicted by the model.

Saylor, D. M., Adidharma, L., Fisher, J. W., & Brown, R. P. (2016). A biokinetic model for nickel released from cardiovascular devices. *Regulatory Toxicology and Pharmacology, 80*, 1-8.

AUTHOR INDEX

Biomedical Engineering Society (BMES)
8201 Corporate Drive
Suite 1125
Landover, MD 20785-2224
USA

ISBN 978-1-5108-6458-0

9 781510 864580

CURRAN ASSOCIATES INC.
proceedings.com